T0341333

Chemical Nanofluids in Enhanced Oil Recovery

Chemical Nanofluids in Enhanced Oil Recovery

Fundamentals and Applications

Rahul Saha, Pankaj Tiwari, and
Ramgopal V.S. Uppaluri

CRC Press is an imprint of the
Taylor & Francis Group, an **informa** business

First edition published 2022
by CRC Press
6000 Broken Sound Parkway NW, Suite 300, Boca Raton, FL 33487-2742

and by CRC Press
2 Park Square, Milton Park, Abingdon, Oxon, OX14 4RN

CRC Press is an imprint of Taylor & Francis Group, LLC

ISBN: 9780367425241 (hbk)
ISBN: 9781032065274 (pbk)
ISBN: 9781003010937 (ebk)

Typeset in Palatino
by codeMantra

Rahul: This book is dedicated to my parents.

Pankaj: To Pragya and Gargi.

Ramgopal: To the Almighty.

Contents

Preface

Continued global emphasis on fossil fuel-based economic growth affirms the prominent role of crude oil in the day-to-day life of the world's population. The globalized economy indicates a continued increase in the crude oil demand and thereby necessitates continued crude oil production profiles at an affordable cost. While oil field explorations have found very few new oil fields, cost-effective recovery of the crude oil from matured reservoirs holds the key for a sustainable energy economy. Thus, pragmatic approaches and technologies are mandatory to facilitate enhanced crude oil production from depleting and about to deplete crude oil reservoirs. Among many alternate technologies, chemical and nanofluid EOR systems have a greater potential for the sustained production of crude oil from depleted reservoirs. This book formalizes the technical and conceptual know-how of the most essential information of the chemical and nanofluid EOR systems.

Acknowledgements

We would like to express our gratitude to all those who helped us in different ways to complete this book. First and foremost, Dr Rahul Saha (first author) wishes to express his deepest gratitude to his parents for their continuous support, guidance and motivation throughout the journey of his life. Their love, courage, care and sacrifice have made it all possible for him to come so far. Furthermore, he wishes to acknowledge the National Institute of Technology Hamirpur for continued motivation to write and complete this book. He is also thankful to all the faculty members, staff and scientific members at the Chemical Engineering Department, Indian Institute of Technology (IIT) Guwahati for their spirited support and cooperation. Furthermore, he is thankful to the analytical laboratory of the Chemical Engineering Department and Central Instrument Facility (CIF) of the IIT Guwahati for allowing him to carry out the scintillating research work in the field of enhanced oil recovery. Last but not least, he wishes to thank his friends, colleagues and lab mates for their active support towards the completion of this book.

Dr Pankaj Tiwari (second author) acknowledges his family, friends, colleagues and of course students for their help in every aspect of his personal and professional life, thereby letting him learn, grow and achieve. He sincerely thanks the Department of Science and Technology (DST), Government of India for funding to initiate the enhanced oil recovery research activities at the IIT Guwahati, and the students who engaged themselves very sincerely with limited resources available during the early stages of his EOR research at IIT Guwahati. The efficient, well-planned and hard work carried out by Dr Rahul Saha and the timely suggestions and path-forward support received from Prof. Ramgopal Uppaluri are highly appreciated and acknowledged. He also wishes to thank his family members, especially his wife Pragya and daughter Utkarsha (Gargi), for their love, care and continued support in his life's journey.

Prof. Ramgopal Uppaluri (third author) wishes to acknowledge the role of the Almighty in his life's journey to reach the wonderful ambience of interdisciplinary research in chemical engineering, materials science, petroleum engineering and food technology. Bestowed with a philosophical outlook, he wishes to acknowledge all those who supported his views and customized thinking in the field of research methodology with rich philosophical treatise and traits, including his family members, his present and past research scholars and friendly colleagues including Prof. Pankaj Tiwari of the Chemical Engineering Department. His good relationship with Prof. Tiwari and Dr Rahul Saha must be sincerely acknowledged for an interdisciplinary endeavour that took the shape in this textbook. Finally, he wishes his most sincere gratitude to the campus of IIT Guwahati that inspires him with its beauty every moment to commit towards high-quality academics and professional research in the field of interdisciplinary subject matter of Chemical Engineering Science and beyond.

Authors

Rahul Saha, PhD, is an Assistant Professor in the Department of Chemical Engineering at the National Institute of Technology Hamirpur, Himachal Pradesh, India. Prior to this position, he served as an Assistant Professor in the School of Petroleum Technology, Pandit Deendayal Petroleum University, Gandhinagar, Gujarat, India. He earned his PhD in Chemical Engineering from the Indian Institute of Technology (IIT) Guwahati, India in 2019. Previously, in 2013, he earned his MTech in Chemical Engineering (Specialization – Petroleum Science & Technology) from the IIT Guwahati, India and BE from Pune University, India. His doctoral thesis focussed on the recovery of crude oil from reservoirs through the application of alternate chemical and nanofluid-enhanced oil recovery schemes. This area of research facilitates pragmatic interdisciplinary research in the broad fields of adsorption kinetics, rheology, colloids and interface science, reservoir engineering and biodiesel production from non-edible oil. His present teaching engagements are devoted to the disciplines of Chemical and Petroleum Engineering.

Pankaj Tiwari, PhD, is an Associate Professor in the Department of Chemical Engineering at the IIT Guwahati. He earned his PhD from the University of Utah, USA in 2012 after carrying out his doctoral research on pyrolysis of complex carbonaceous complex solid material, oil shale. He also worked at the General Electric, Plastic division at JFWTC Bangalore on developing the monomer for high-performance polymer (HPP). His continued research activities have been devoted to gaining useful insights into alternative and competent mechanisms of enhanced oil recovery and the development of kinetics and compositional models for pyrolysis and combustion processes. To date, he has supervised five PhD students and published more than 35 international journal publications.

Ramgopal V.S. Uppaluri, PhD, earned his PhD in Process Integration from the University of Manchester, UK and MTech in Chemical Engineering from the IIT, Kanpur. For more than a decade, he has been a Faculty of Chemical Engineering Department at IIT Guwahati. His research interests are varied and include membrane technology, petroleum engineering and refining, food engineering and processing, process optimization, wastewater treatment, bio-fertilizers, evolutionary engineering optimization and so on. To date, he has supervised 13 PhD students in diverse fields of process engineering and published 110 international journal publications. In upstream engineering, his research is primarily in the field of chemical and polymer-enhanced oil recovery.

Abbreviations

AMPS	2-acrylamido-2-methylpropanesulfonic acid
API	American Petroleum Institute
ASP	Alkali surfactant polymer
BET	Brunauer–Emmett–Teller
CF	Chemical flooding
CMC	Critical micelle concentration
CMHPG	Carboxylmethyl hydroxyl propyl guar
CMHEC	Carboxymethyl hydroxyl ethyl cellulose
CSS	Cyclic steam stimulation
CTAB	Cetyltrimethylammonium bromide
DBS	Dodecylbenzenesulfonate
EDX	Energy dispersive X-ray
EOR	Enhanced oil recovery
FESEM	Field emission scanning electron microscope
FTIR	Fourier transform infrared spectroscopy
HBL	Hydrophilic–lipophilic balance
HPAM	Partially hydrolyzed polyacrylamide
HPG	Hydroxyl propyl guar
IFT	Interfacial tension
IIOP	Initial oil in place
IOR	Improved oil recovery
LS	Lignosulfonate
NF	Nanofluid flooding
NMR	Nuclear magnetic resonance
OOIP	Original oil in place
OPEC	Organization of the petroleum exporting countries
PHPA	Partially hydrolyzed polyacrylamide
PONP	Nonylphenyl polyoxyethylene glycol
ppm	Parts per million
PS_20	Polysorbate20
PV	Pore volume
PVA	Vinyl phosphonate
SAC	Strong acid cation
SAGD	Steam-assisted gravity drainage
SDBS	Sodium dodecylbenzenesulfonate
SDS	Sodium dodecylsulfate
SNPs	Silica nanoparticles
TX-100	Triton X-100
TGA	Thermogravimetric analysis
VES	Viscoelastic surfactant

WAC	Weak acid cation
WF	Water flooding
XRD	X-ray powder diffraction

Symbols

E	Overall hydrocarbon displacement efficiency
E_V	Microscopic hydrocarbon displacement efficiency
E_D	Macroscopic hydrocarbon displacement efficiency
M	Mobility ratio
N.C.	Not critical
O/W	Oil in water
W/O	Water in oil
σ	Interfacial tension (mN/m)
ρ_H	Density of heavy water phase (g/cm^3)
ρ_L	Density of light oil phase (g/cm^3)
λ_o	Mobility of oil
λ_w	Mobility of water
ω	Rotation (rpm)
k_{ro}	Relative permeability of oil
k_{rw}	Relative permeability of water
μ_o	Viscosity of oil
μ_w	Viscosity of water
N_c	Capillary number
S_{or}	Residual oil saturation
Δp	pressure drop
ΔH	Enthalpy (kJ/mol)
μ	viscosity
ν	Darcy velocity
L	length (mm)
D	diameter (mm)

Introduction

This book elaborates tertiary chemical-enhanced oil recovery (EOR) schemes. Such schemes have been concluded to have a great potential for the cost-effective recovery of residual oil from reservoirs being subjected to primary and secondary flooding operations. The central goal of the book is to understand and visualize various complex reaction mechanisms associated with chemical and nanofluid EOR systems. These include generic insights associated with alkali, surfactant, polymer and nanofluid flooding systems. The potential roles of alternate reservoir system parameters such as formation water salinity, reservoir temperature, crude oil properties, porosity and reservoir rock permeability have been explored to classify various possible combinations for the successful deployment of the applicable chemical EOR system. Pertinent responses during EOR include emulsification extent, IFT reduction, injection pattern specific recovery, neutralization and saponification, wettability alteration, thermal stability and so on. Thereby, screening criteria have been summarized to serve as a useful guideline with rules of thumb to assist practising engineers and consultants in the field of enhanced oil recovery. A successful enterprise of the chemical EOR is always related to the market price of the crude oil commodity. Nonetheless, deployed chemicals incur additional costs per unit barrel of crude oil produced. Technically, these related to the efficient utilization of surfactants and other chemicals that perform outstandingly in terms of poor surfactant adsorption characteristics of the rock surface and poor degradation characteristics under pertinent reservoir system conditions. Thus, surfactant adsorption kinetics have been elaborated in the volume based on various isotherm models. Furthermore, investigations associated with thermal stability have been discussed. Finally, the book emphasizes the application of chemical nanofluid EOR methods in conventional and unconventional reservoirs. The nanofluid EOR system is an emerging technology in the oil and gas industry and a technical knowledge of the same can provide useful dividends for those who wish to apply such sophisticated technology in crude oil production schemes.

In summary, the overall purpose and goal of this book are to disseminate useful information with respect to the technical know-how of alternate chemical and nanofluid EOR systems and thereby serve as a useful guide for practising engineers and consultants. Thereby, the volume aims to converge upon the professional activities of such skilled personnel towards the definitive application of chemical and nanofluid EOR in field applications and paves the way for a fossil fuel-based sustainable world economy.

1

Introduction to Chemical and Nanofluids-Induced Oil Recovery

1.1 Importance of Crude Oil

The world economy after globalization has witnessed a massive transformation in terms of industrialization, transportation and urbanization. Invariably, a greater proportion of such advancement involves the overwhelming consumption of fossil fuels such as crude oil which are the gifts provided to mankind by mother earth. Constituting a mixture of hydrocarbons, crude oil is a naturally occurring thick dark brown flammable liquid. Crude oil is hypothesized to have formed due to the decomposition of dried plants and animals under the prevalent high temperature and pressure conditions beneath the earth's surface that existed millions of years ago.

Ancient civilizations did not have the know-how of crude oil, and they very much depended upon tar in the Stone Age to make brick and mortar to enable their residences to be waterproof. Additionally, tar-based coatings were used during these civilizations for the leak-proof requirements of their boats. Contrary to this, today, a large portion of the growing world energy demands is being fairly met with the adequate production and supply of crude oil. Moreover, current modern society is the major consumer of a wide variety of crude oil-based by-products. The huge crude oil reserves of the Middle Eastern countries transformed them into flourishing rich economies undergoing rapid urbanization. Hence, crude oil is regarded as black gold and extensively catalyses the economic growth of many countries. Worldwide market demand and supply dictate the fluctuations and stabilities of crude oil prices and are coordinated by the Organization of the Petroleum Exporting Countries (OPEC).

1.2 Crude Oil – Demand and Supply

The existing energy consumption profiles in the world indicate the near impossibility of replacing crude oil completely with renewable energy. As a

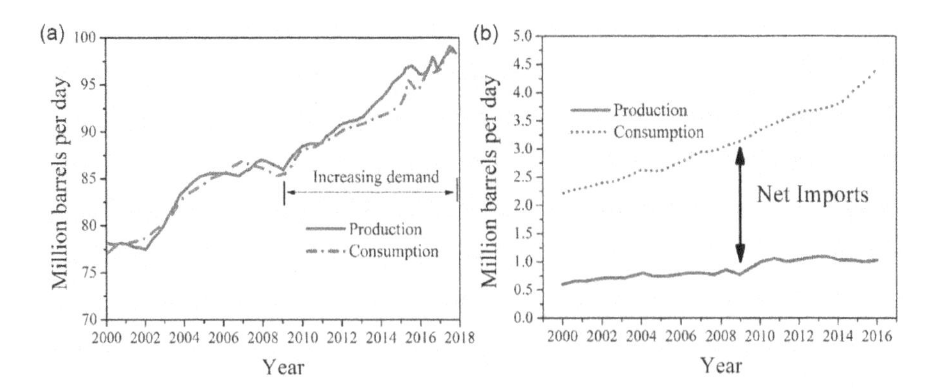

FIGURE 1.1
Petroleum liquid production and consumption in (a) worldwide and (b) India scenario [1].

major contributor to meet the world's energy demands, crude oil contributes enormously to the world economy. Figure 1.1a depicts an increase in oil production and consumption profiles on a worldwide basis. As depicted in the figure, during the initial time period 2000–2009, the net crude oil production and consumption in the world increased from 77 million to 85 million barrels per day. However, in the past 10 years, i.e., from 2009 to 2019, the net world crude oil production and consumption rate rose sharply from 85 million to 98 million barrels per day. This is significantly higher in comparison to the previous 10 years' time period and thus indicates higher demand for crude oil.

For India, the profiles are distinct as shown in Figure 1.1b. Being one of the largest importers of crude oil, India's net import is bound to increase in the near future, given that the trend grew from 73% in 2000 to 77% in 2016. Despite the improved oil production rate from 0.75 (2009) to 1 million barrels (2016), the crude oil import rate has increased in recent years at a greater rate in comparison to that prevalent in the country prior to 2010. In summary, both figures are in good agreement with one another to infer upon the fact that the worldwide crude oil demand and crude oil import rate for India are increasing significantly after 2009.

The ongoing demands for crude oil are mainly due to continuous increase in energy consumption and the use of hydrocarbon-based products by modern society. Thus, a greater emphasis is placed on the cost-effective production of larger quantities of crude oil either from newly discovered reservoirs or from mature reservoirs that are already operating past their peak production levels. This is because the industry does not guarantee the discovery of new oil reserves, and if they are discovered, it is likely they will be located deep offshore or in other areas difficult to access. Also, the crude oil production from unconventional sources will be very expensive from mature oil fields [2]. Hence, considering the available data for Indian crude oil reservoirs, mature oil reservoirs have been targeted for enhanced oil production. Figure 1.2 represents the distribution of crude oil reserves in

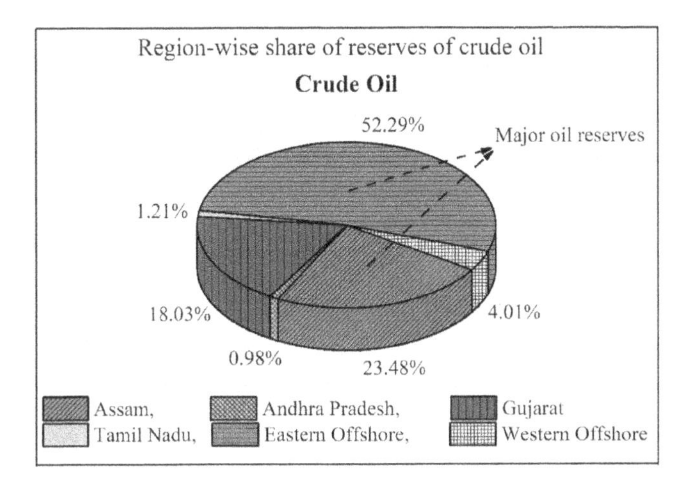

FIGURE 1.2
Region-wise reserves of crude oil in India [3].

India. The chart illustrates that the majority of oil reserves exist in the eastern offshore (52.29%) and Assam regions (23.48%) of India. Hence, it is apparent that to increase crude oil production in such regions of India and to meet the growing energy demands of the country, enhanced oil recovery methods are to be deployed. Thereby, net import costs associated with the crude oil system can be minimized along with an emphasis on the self-sustainability of the country's demanding energy needs.

1.3 Enhanced and Improved Oil Recovery

The recovery of crude oil from reservoirs is accomplished due to the natural pressure difference that exists within the bottom well hole and the oil field. Oil recovery by such natural pressure differences is known as the primary method. However, with continued oil production, the pressure inside the reservoir system decreases. Therefore, to overcome this feature, the pressure inside the reservoir system is increased through the injection of water during secondary flooding operations. It is well known that the oil production from primary and secondary (water) flooding operations accounts for only 30%–40% of the original oil in place (OOIP) [4,5]. Thus, even after secondary recovery, about two-thirds of the OOIP is not recovered and remains trapped in the porous structure of the reservoir rock [5,6]. The viscous fingering effect of water flooding is primarily responsible for trapping crude oil in such a system. Therefore, more complex procedures and methodologies need to be adopted to recover the trapped oil economically. The enhanced

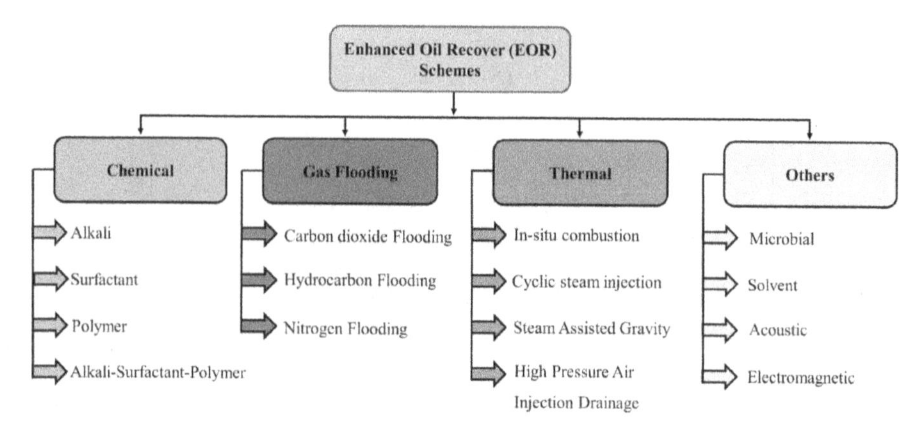

FIGURE 1.3
Schematic of various enhanced oil recovery methods.

oil recovery (EOR) process is a tertiary oil recovery process that deploys advanced methods for oil extraction. Depending on the working principle fundamentally explored, the EOR schemes are often classified into various categories such as thermal, gas, chemical and other recovery processes (Figure 1.3). On the other hand, the improved oil recovery (IOR) method involves a wider production range of recovery technologies. Thus, EOR processes are regarded as sub-classes of the IOR methodology [2]. A brief of various EOR methods is presented as follows.

1.3.1 Thermal-Enhanced Oil Recovery

This method involves heating of the crude oil. The thermal EOR targets the application of either heat or thermal energy to the reservoir system, which reduces the viscosity of crude oil and enhances oil recovery [7–9]. The well-known thermal EOR methods include hot water or steam injection, stimulation using cyclic steam, combustion (*in-situ*) and gravity drainage assisted with steam. The steam-based methods involve the injection of steam into the reservoir through injection wells. Therefore, crude oil viscosity is effectively reduced and aids in the oil's flow to the surface of the reservoir production wells. The *in-situ* combustion involves generation of heat through burning a portion of crude oil in the reservoir itself. This is achieved through the injection of air or oxygen-enriched air. Eventually, oil recovery is enhanced through the generated heat and gases from combustion. Thermal recovery methods are applicable for low-depth reservoirs with highly viscous oil and lower API [10]. Thermal recovery methods constitute the risk of safety issues during larger production schemes and can severely damage the underground oil well structure. This is regarded to be a very serious limitation of the thermal EOR technique in oil reservoirs [11].

1.3.2 Gas-Enhanced Oil Recovery

The gas injection method involves injection of miscible and immiscible gases into the reservoir and is known to be one of the oldest EOR techniques. The injected gas dissolves into the oil phase and eventually reduces crude oil viscosity and interfacial/surface tension of the oil–water interface. Thereby, sweep efficiency can be improved considerably to enhance the production of crude oil. Among alternate gas injection schemes, nitrogen and flue-gas injection, hydrocarbon injection, CO_2 flooding and so on are popular for their precise screening criteria in flooding applications [10]. For nitrogen and flue-gas flooding, these correspond to a depth of >6000 ft and an API gravity of 35°–48°. For hydrocarbon injection, the screening criteria refer to a depth of >4000 ft and an API gravity of 23°–41°. For the CO_2 flooding method, the suggested depth is more than 2500 ft and the recommended API gravity is about 22°–36°. The CO_2 flooding method is also applicable at higher depths and higher API gravities. Additionally, during gas flooding specific reservoir conditions are required which can expand the injected gas and drive the crude oil for recovery [11].

1.3.3 Chemical-Enhanced Oil Recovery

The chemical EOR process involves the recovery of residual oil through the injection of chemicals into the reservoir system. The injected chemicals interact with the crude oil and enable the alternation of pertinent mechanisms to enhance oil recovery. The chemical EOR system performance is very much dependent upon the characteristics of the reservoir system such as crude category, rock properties, temperature and salinity.

Apart from chemical EOR, other EOR categories such as microbial, solvent, acoustic and electromagnetic technologies do exist. However, a detailed account of the working principle and relevant screening criteria is beyond the purview of this book.

1.4 Chemical-Enhanced Oil Recovery Mechanisms

Among various EOR methods, the chemical flooding technique is an important residual oil recovery production technology. It is carried out after the secondary water flooding to recover the residual oil prevalent in the crude oil reservoir. Chemical EOR is usually targeted with alkali, surfactant and polymer or their combinations to increase capillary number, reduce interfacial tension, emulsify the crude oil, improve overall oil displacement efficiency, facilitate mobility control and alter wettability [10,12–21]. Recently, the application of nanoparticles in chemical EOR has been investigated, but as per our knowledge, there are no filed applications as it is in the development stage. Hence, this book is designed to focus on the applications of chemical-nanofluid EOR in addition to chemical EOR. The important dimensionless number and mechanisms associated with chemical EOR applications are described in the section below.

1.4.1 Capillary Number

A significant amount of residual oil gets entrapped in the reservoir even after conducting water flooding operations. This trapping of the crude oil in the porous structure of the reservoir is due to the lower capillary number of the system, which is a function of the pertinent capillary forces. Defined as a dimensionless number, the capillary number (N_c) is expressed as the ratio of viscous forces to capillary forces [22], i.e.,

$$N_c = \frac{\mu v}{\sigma} = \frac{k\Delta p}{\sigma L} \tag{1.1}$$

where μ, v, σ, k and $\frac{\Delta p}{L}$ correspond to displacing fluid (water) viscosity, Darcy velocity, interfacial tension (IFT) between oil and water, effective permeability (of displacing fluid) and pressure gradient across the length, respectively.

Hence, oil mobility is directly influenced by the capillary number. The higher the capillary number, the higher the oil mobility and oil recovery. This can be accomplished through the simultaneous reduction in the IFT of the oil–water system and enhancement in the viscosity of the displacing fluid. The chemical flooding operations target enhancement in the capillary number through the reduction of IFT or enhancement in the viscosity of the displacing fluid. Figure 1.4 depicts the variation of normalized residual saturation with the capillary number for sandstone cores. As depicted, the residual saturation reduced with an increase in the capillary number.

1.4.2 Interfacial Tension (IFT)

Consider the interface associated with a system of two immiscible systems. For such a system, the IFT is defined as the force required per unit length to create a new surface area at the said interface. In other words, IFT refers to the force per unit length that is to be applied parallel to the surface to create a new surface area. Often the applied force is perpendicular to the concentration gradient or local density of the system.

IFT is an interfacial phenomenon, and for a crude oil-chemical system, the IFT characteristics (being expressed in terms of either equilibrium or dynamic IFT values) are dependent upon the interactions associated with the prevalent crude oil and chemical phases at the interface. The IFT of crude oil-aqueous chemical phase system can be measured using a spinning drop

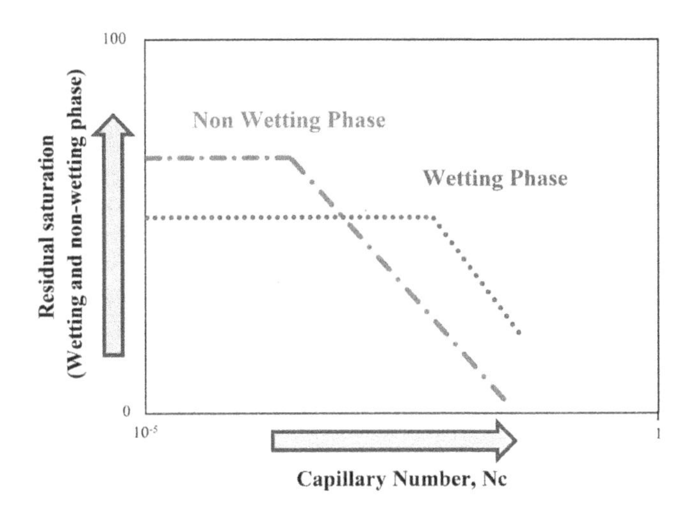

FIGURE 1.4
Relation between capillary number and residual oil saturation for the wetting and non-wetting phases in sandstone cores [12]. (Modified from Lake, L. (1989). Enhanced Oil Recovery. E. Cliffs. New Jersey, Prentice Hall.)

tensiometer. A typical spinning drop tensiometer measures the IFT of the crude oil system by using the following expression:

$$\sigma = \frac{r^3\omega^2(\rho_H - \rho_L)}{4}; \frac{L}{D} \geq 4 \tag{1.2}$$

where σ, ρ_H, ρ_L, ω, D ($2r$) and L are the IFT (mN/m), heavy water phase density (g/cm^3), light oil phase density (g/cm^3), velocity of the spin or rotation (rpm), width of the measured drop (mm) and oil drop length (mm), respectively. A reduction in IFT enables the desired release of trapped residual oil and therefore enhances crude oil recovery. Due to the strong relevance of chemical characteristics in the system, the IFT is influenced by temperature and system salinity.

1.4.3 Emulsification

The emulsification process involves effective mixing of two immiscible liquids through an emulsifying agent. The agent facilitates effective dispersion of one phase into another phase to facilitate the formation of droplets. For crude oil systems, emulsions have been categorized into either O/W (oil in water) and W/O (water in oil) emulsions. For the O/W emulsion case, oil gets dispersed in water. However, for the O/W emulsion, water gets dispersed in oil. The chemical EOR processes utilize surfactants that possess a water-soluble hydrophilic head and an oil-soluble hydrophobic tail as emulsification agents. The hydrophilic–lipophilic balance (HBL) factor accounts for the degree of hydrophilicity or lipophilicity of the surfactant. Typically, a surfactant possessing a 3–6 HBL value confirms the achievement of W/O emulsions. On the other hand, surfactants possessing 8–16 HBL would favour the achievement of O/W emulsions.

In a typical chemical EOR system, phase behaviour investigations are usually carried out to identify the most suitable surfactant. In due course of such analysis, three different types of micro-emulsions, namely lower (type I), middle (type III) and upper (type II) emulsions, can be observed which are classified as Windsor type emulsions. Among these, the type III region is regarded to be the optimal micro-emulsion due to the existence of all three phases (oil, micro-emulsion and water) in equilibrium. Such a three-phase region also favours the achievement of ultra-low IFT values [13,23].

1.4.4 Displacement Efficiency

The reservoir volume in contact with the injected liquid directly influences the displacement-based oil recovery characteristics. The overall displacement efficiency of hydrocarbons is dependent upon two factors, namely, microscopic efficiency and macroscopic efficiency, i.e.,

$$E = E_V * E_D \tag{1.3}$$

where E, E_V and E_D are overall, microscopic and macroscopic hydrocarbon displacement efficiencies, respectively.

Among these efficiencies, the microscopic displacement efficiency refers to the oil mobilization or displacement at the level of the pore structure of the reservoir rock. Thereby, the factor enables the measurement of the displacing fluid efficacy to displace the oil existing within the pores of the rock. On the other hand, the macroscopic or volumetric sweep efficiency refers to the efficacy of the displacing fluid to sweep out the reservoir volume in both aerial and vertical directions. Therefore, this efficiency refers to the ability of the displacing fluid to move out the dispersed oil to reach the production wells.

1.4.5 Mobility Ratio

It is defined as the ratio of the displacing fluid mobility to displaced fluid/oil mobility. The mobility ratio (M) during a waterflood operation is defined as follows:

$$M = \frac{\text{Mobility of water}}{\text{Mobility of oil}} = \frac{\lambda_w}{\lambda_o} = \frac{k_{rw}/\mu_w}{k_{ro}/\mu_o} = \frac{k_{rw}\mu_o}{k_{ro}\mu_w} \tag{1.4}$$

where λ_w and λ_o are mobilities of water and oil (mD/cp), respectively; k_{rw} and k_{ro} are water and oil relative permeabilities, respectively; μ_o and μ_w are oil and water viscosities, respectively.

Several interesting scenarios exist with respect to the values of *M*. For the system in which the entering fluid bypasses the displaced fluid, the value of *M* will be >1. Such conditions are not desired due to the very fact that they indicate poor sweep efficiency and hence limited oil recovery. To achieve the highest combinations of displacement efficiency and oil recovery, the M value should be ≤1. For all such cases, water viscosity enhances and oil viscosity decreases.

1.4.6 Wettability Alteration

It is the fluid ability to maintain contact or adherence to the surface of a solid in the presence of other immiscible fluids and is an important surface principle in the chemical EOR system. An alteration in the wettability enables the oil to flow to the production platform through the needful replacement of trapped oil in the pore space with water. Wettability of a reservoir system is a highly complex parameter. It is strongly influenced by rock mineralogy, oil/water composition, initial water saturation and temperature.

It is well known that oil-wet reservoirs do not facilitate higher oil recovery due to the inability to imbibe water into the rock matrix during flooding operations. However, using a chemical slug, it is possible to alter the wettability of the reservoir system from oil-wet to water-wet characteristics and thereby enhance the oil recovery factor. Hence, wettability alteration is a vital mechanism in chemical EOR and shall not be ignored while performing laboratory, pilot and field-scale experiments.

1.5 Selection Criteria for EOR and Chemical EOR Processes

Extensive investigations in the United States enabled the development of adequate rules of thumb focusing on the estimation of crude oil range and reservoir properties. Several factors influence and converge upon the appropriate selection criteria associated with the applicability of EOR to a specific combination of reservoir, crude and rock, characteristics. Table 1.1 elaborates upon this information in a concise format [10,13].

Due to the associated complexity of the chemical EOR processes, standard selection criteria could not be easily formulated. Such criteria significantly depend upon the brine formation, oil, formation rock, reservoir temperature and rock permeability characteristics [2,13]. The prevalent divalent ions such as Ca^{++} and Mg^{++} in the formation of a water system contribute towards the brine hardness characteristics. Therefore, the interaction of such constituents with the induced chemicals would facilitate the formation of precipitates that severely detriment the stability of the induced chemical solutions. A desirable characteristic of the induced chemical EOR system is that it should be water/

TABLE 1.1
Screening criteria summary for EOR methods [10, 13, 28]

EOR Method	Oil Properties			Reservoir Properties					
	Gravity °API	Viscosity (cp)	Composition	Formation Type	Net Thickness (ft)	Depth (ft)	Average Permeability (md)	Oil Saturation (%PV)	Temperature (°F)
Chemical EOR									
Micellar polymer, ASP, and alkali flooding	>20	<35	Light, intermediate, some organic acids for alkaline floods	Sandstone preferred	N.C.	<9,000	>10	>35	<200
Polymer flooding	>15	<150 (>10 and <100 preferable)	N.C.	Sandstone preferred	N.C.	<9,000	>10	>50	<200
Thermal EOR									
Steam	8 – 25	<100,000	N.C.	Sandstone with high porosity and permeability preferred	>20	<5,000	>200	>40	N.C.
In-situ combustion	10 – 27	<5,000	Some asphaltic components	High porosity sand or sandstone	>10	<11,500	>50	>50	>100

(Continued)

TABLE 1.1 (*Continued*)

Screening criteria summary for EOR methods [10, 13, 28]

EOR Method	Oil Properties			Reservoir Properties					
	Gravity °API	Viscosity (cp)	Composition	Formation Type	Net Thickness (ft)	Depth (ft)	Average Permeability (md)	Oil Saturation (%PV)	Temperature (°F)
Gas Injection Methods (Miscible) EOR									
Hydrocarbon	>23	<3	Light hydrocarbon – high%	Sandstone or carbonate	Thin unless formation is dipping	>4,000	N.C.	>30	N.C.
Nitrogen and flue gas	>35	<0.4	Light hydrocarbon – high%	Sandstone or carbonate	Thin unless formation is dipping	>6,000	N.C.	>40	N.C.
Carbon dioxide	>22	<10	Intermediate hydrocarbon (C_5–C_{12}) – high%	Sandstone or carbonate	Wide range	>2,500	N.C.	>20	N.C.
Immiscible gases	>12	<600	N.C.	N.C.	N.C. if dipping and/or good vertical permeability	>1,800	N.C.	>35	N.C.

Source: (Modified from Taber, J. J., et al. (1997). "EOR Screening Criteria Revisited - Part1: Introduction to Screening Criteria and Enhanced Recovery Field Projects." SPE Reservoir Engineering 12: 189-198.)

oil soluble and shall interact with the crude oil system to enable a significant reduction in the interfacial tension (to reach ultra-low values). The achievement of ultra-low IFT can also be investigated by conducting phase behaviour investigations [13,24]. The corresponding emulsions being miscible phase shall propagate through the porous structure of the reservoir rock system. To a significant extent, viscosity and emulsion stability govern microscopic and macroscopic displacement efficiency [20,25–27]. Moreover, the chemical slug selection is highly dependent upon the category of the crude oil (characterized as light, moderate or heavy crude oil). Furthermore, chemical adsorption is inevitable on the reservoir rock surface. These adsorption characteristics are highly dependent upon the charge characteristic of the surfactant and rock system. In general, an anionic surfactant for carbonate reservoirs and vice versa is not recommended due to enhanced adsorption characteristics encountered by opposite charge interaction. Moreover, in carbonate reservoirs, the anhydrites present in the carbonates enable precipitation and thereby contribute towards higher consumption of the alkali. Furthermore, clay enables higher chemical adsorption, and its constitution in the rock should be low for minimal adsorption and successful chemical EOR [2]. The prevalent extreme temperatures in the reservoir system can deteriorate chemical slugs to thereby render them ineffective towards the very purpose of chemical flooding. Also, the low permeability of the reservoirs translates into issues associated with injection procedures and excess retention of chemicals. Therefore, higher permeability of the reservoir systems is highly desirable for effective application of chemical flooding systems. Furthermore, the identified combinations of chemicals and their formulation for a particular oil well should activate dominant mechanisms associated with higher oil recovery.

1.6 Overview of Chemical and Chemical-Nanofluid EOR

A combination of chemical and chemical-nanofluid flooding systems can be deployed in a certain reservoir after duly analysing the net effect of crude oil, reservoir properties, temperature and saline water characteristics, and the favourability of a particular mechanism [20,28–38]. We know that chemical EOR is a process that involves the application of different chemicals like alkalis, surfactants, polymers and their combination to enhance oil recovery. In the chemical-nanofluid system, a stable aqueous chemical solution is being formulated using various chemicals and nanoparticles. The implementation of such aqueous nanofluid solutions can target the residual oil recovery from heterogeneous oil reservoirs. The overview of different chemical and chemical-nanofluid injection schemes is explained below:

Alkali flooding – In several countries (USA, Canada, China and Saudi Arabia), alkali flooding was primary targeted because of its simplicity and lowest cost-efficacy as compared to various alternate chemical solution systems [14,15,27,39–41]. In a typical chemical EOR system, the interfacial tension is reduced by targeting the generation of *in-situ* surfactants at the oil-water interface through the reaction of alkali with the acidic constituents of crude oil [42,43]. Therefore, due to such chemical EOR mechanism, researchers invariably targeted to accomplish a suitable correlation among crude oil acid value, IFT reduction and residual oil recovery.

Cooke et al. [15] observed that crude oil with an acid value of over 1.5 favours alkali flooding through the formation of an oil emulsion bank. The authors conveyed that both IFT reduction and emulsification are responsible for higher oil recovery. Similar studies using four different oil reservoirs (Zhuangxi 106, Chenzhuang, Binnan and Xia-8) with different acid values (1.846, 2.018, 3.852 and 4.660 mg of KOH/g of sample, respectively) were investigated to examine the influence of acid value on the recovery of residual oil [25]. The data depicted that the oil recovery factor has increased from 12.4% to 20.4% with an increase in acid value from 1.846 to 4.660 mg of KOH/g of the sample. The crude oil with a higher acid value is known to possess higher organic acid content and facilitates higher enrichment in terms of *in-situ* surfactants. Instantaneous IFT reduction is favoured by the presence of *in-situ* surfactants at the O/W interface, which improves penetration of alkali solution in the oil phase to form W/O emulsions. Such dispersion promotes sweep efficiency by reducing both displacing fluid viscosity and viscous fingering effect. Therefore, the investigations carried out infer that IFT reduction is not the only criteria for successful alkali flooding but the emulsification mechanism also contributes significantly towards the oil recovery factor. Similar views have also been reported by other researchers to confirm that the emulsification mechanism contributes significantly towards the oil recovery factor [39,40,44]. Additionally, researchers have elaborated upon the alkali (NaOH and Na_2CO_3) effect on the variations in wettability alteration of the reservoir rock [45]. The authors found that wettability alteration improved displacement efficiency and increased residual oil recovery by about 10%. The micro-model experiments conducted using an alkali solution clearly indicated a favourable alteration in pore wall wettability from water-wet to favourable oil-wet. However, no direct relationships among acid value, equilibrium IFT, emulsification, alteration in wettability and residual oil recovery have been observed to affirm a successful chemical EOR process [19,46]. Recent studies revealed that the displacement of oil from pores of reservoir rocks involves the phenomena to occur at the oil–water interface (IFT reduction) and the oil–solid/rock interface (wettability) [47,48].

Surfactant flooding – Several investigations targeted the efficacy of surfactant flooding to achieve residual oil recovery from crude oil reservoir systems [49–52]. As early as 1960, surfactant injection was considered to enhance oil recovery from the reservoirs [53]. Surfactant flooding enhances sweep efficiency to eventually facilitate a reduction in IFT, alteration in wettability and formation of emulsions and hence improved oil recovery [51,54–57].

Therefore, due to these combined effects, surfactant flooding has been identified as an efficient and effective method to successfully recovery residual oil from crude oil reservoirs. In such investigations, various surfactants considered till date include anionic (alkyl ether sulfate [51], sodium lauryl sulfate [31], sodium dodecyl benzene sulfonate [33] and so on), cationic (hexadecyltrimethylammonium bromide [HTAB] [31], cetyltrimethylammonium bromide [CTAB] [56] and so on), non-ionic (Tergitol [31] and so on) and natural surfactants [32,58].

Surfactant screening is required in the context of negligible degradation in high-temperature saline crude oil reservoirs. Targeting the application of surfactants in such extreme reservoir conditions, investigations were carried out to evaluate the thermal ageing of two different surfactants [59]. The authors targeted characterizations of thermally aged surfactants using FTIR, NMR and TGA [54]. Thereby, the influence of thermal ageing on the prevalent functional groups has been addressed. Both thermally aged and non-aged samples were studied in terms of IFT reduction characteristics. The thermogravimetric analysis (TGA) inferred upon the short-term thermal behaviour of the considered material. Such analysis in the carried out investigations confirmed no signs of degradation for both surfactants A (carboxybetaine-based amphoteric surfactant) and B (anionic surfactant Alfoterra L167-4s). However, after thermal ageing of both samples at 90°C for 10 days, the FTIR and NMR spectra confirmed that the surfactant B underwent degradation. For the amphoteric surfactant, it is important to note that its IFT characteristics did not deteriorate and vary with thermal ageing. However, for the anionic surfactant, the IFT value increased by two orders of magnitude after ageing. The stable amphoteric surfactant was further investigated to evaluate the influence of temperature and system salinity. In such investigations, the authors analysed that while system IFT enhanced with increasing temperature, it reduced with increasing salinity.

Studies on 12 surfactants from 5 different categories were examined for effective surfactant flooding to enhanced residual oil recovery [20]. It explored the extreme reservoir conditions by varying the temperature (90°C–120°C) for a fixed choice of salinity (20×10^4 mg/L). For these process parametric considerations and for an ageing period of 125 days, two surfactant formulations indicated superior performance with the achievement of very low values of IFT ($\leq 10^{-3}$ mN/m). It was analysed that with a reduction in initial and dynamic IFT, the oil recovery was enhanced and an additional oil recovery of 7% was achieved through the reduction in IFT from 10^{-1} to 10^{-3} mN/m. However, for an IFT reduction of up to 10^{-4} mN/m, the oil recovery and IFT reduction correlation could not be validated as under such a situation, the emulsification process played a dominant role. The emulsification condition maximizes the oil recovery to 36.65% with an enhanced surfactant concentration from 0.2 to 0.5 wt%. The emulsification became too strong for a variation in surfactant concentration from 1 to 3 wt% and under such scenario, the displacement process was severely affected. Hence, despite achieving ultra-low

IFT values for the chosen surfactant concentrations, the oil recovery reduced but did not enhance. Researchers conducted dilute surfactant flooding experiments to investigate the role of IFT reduction and emulsification on the displacement efficiency of the system [60]. The investigations involved fixed surfactant concentration (0.2 wt%) and salinity variation to obtain solution systems possessing various levels of IFT (10^{-1} to 10^{-5}mN/m). Such formulations have been considered by the authors to eliminate the influence of surfactant concentration on the displacement efficiency characteristics. For the said range of IFT levels (10^{-1} to 10^{-5}mN/m) that were achieved for variant surfactant concentrations, the displacement efficiency varied from 12.44% to 21.45%. In other words, the IFT reduction did influence the displacement efficiency values significantly. However, it is important to note that the impact of emulsification is on the contrary, i.e., the displacement efficiency reduced with increasing emulsification. Thus, the best emulsification characteristics existed for the system with lower IFT.

Using cationic (C_{12}TAB), anionic (sodium dodecylsulfate – SDS) and non-ionic (Triton X-100) surfactants, a mechanistic investigation upon the alternations of wettability for the carbonate rock (dolomite) system was examined [57]. Among the chosen surfactants, C_{12}TAB was found to be highly efficient to facilitate an alteration in the wettability to enhance water wet characteristics. This is due to the role of ionic interaction to facilitate irreversible desorption of stearic acid from the surface of the carbonate rock system. On the other hand, for the Triton X-100 case, both ion exchange and π electrons polarization contributed towards its adsorption onto the surface of the dolomite simultaneously releasing stearic acid from the dolomite surface. The adsorption of this releases stearic acid forming a new layer on the surface resulted from the hydrophobic interaction, which eventually resulted in a weak water-wet system. SDS indicated neutral wet condition for the chosen system as adsorption of SDS on the carbonate rock sample was due to the hydrophobic interaction of adsorbed acid and surfactant tail. The order of surfactant efficiency in terms of wettability alteration was found to be C_{12}TAB > Triton X-100 > SDS.

The adsorption of surfactant within an oil reservoir is another important parameter that needs to be evaluated during a chemical EOR process. The adsorption characteristics of surfactants are dependent on rock characteristics. Hence, appropriate analysis is required prior to injecting surfactants into crude oil reservoirs for EOR applications. The authors carried out fundamental investigations with respect to the determination of interfacial properties and adsorption characteristics of four different surfactants (two anionic – DBS and LS, and two non-ionic – PS20 and PONP) on kaolinite samples [61]. The Langmuir and Freundlich isotherm models have been evaluated to fit well with measured equilibrium adsorption data. However, the former model provided better fitness. The adsorption of various surfactants has been in the following order: LS > PS20 > DBS > PONP (with adsorption capacities of 5.28, 3.45, 3.26 and 3.06 mg/g, respectively). For 1 wt% surfactant solution, the IFT reduction of model oil-surfactant solution system has been evaluated to be as per the following order: DBS>PONOP>PS20>LS. Based on IFT reduction behaviour and financial loss incurred due to surfactant loss, the authors concluded that DBS is the best surfactant among all four considered surfactants. Another study investigated the effect of rock mineralogy on surfactant adsorption using SDS (anionic) and Triton X-100 (non-ionic) surfactants by the CMC method [62]. Mineralogical investigations referred to the mixing of quartz, kaolinite, feldspar, illite and montmorillonite in various proportions to evaluate the adsorption characteristics. For Triton X-100, the adsorption was highest with montmorillonite and the adsorption capacity increased from 1.5 to 28.5 mg/g for an increase in the montmorillonite concentration from 5% to 20%. For an anionic surfactant, adsorption was negligible and is due to the nature of the charge associated with both minerals and surfactant. Therefore, based on adsorption characteristics, the authors inferred that the mineralogical order for a non-ionic surfactant is as per the following order: montmorillonite>illite>kaolinite. Researchers also used natural surfactants for EOR application and further investigated their adsorption characteristics. The adsorptive behaviour of *Ziziphus spina-christi*, a non-ionic natural surfactant with the carbonate rock system, was investigated [32]. The adsorption behaviour was operated within the temperature range of 28°C–75°C. An exothermic adsorptive behaviour was apparent for the system that affirmed reduction in the adsorptive quantity of surfactant with increasing temperature. The measured adsorption data have been evaluated to best fit with the Freundlich equilibrium adsorption isotherm and the pseudo-second-order kinetic model (R^2 values obtained were >0.99 for both cases). The CMC of the natural surfactant was obtained as 3.65 wt%, and the IFT reduction encountered for such a system was around 69%. The IFT reduction by the natural surfactant increased imbibing into the crude oil system to significantly enhance the oil recovery factor from 47% to 77%.

Polymer flooding – Polymer flooding is adopted for heavy crude oil systems due to the very reason that the polymer facilitates the maintenance of a favourable mobility ratio and hence better sweep efficiency. This is not the case for water flooding-based secondary recovery operations. In such scenarios, poor recovery factors are often attributed to poor sweep efficiency and viscous fingering effect. Therefore, to overcome these issues, polymer flooding has been suggested as early as 1950 [17,63].

The inclusion of a polymer in aqueous media enables thickening effects and thus alters the rheological characteristics of the system. Therefore, the water channelling effect is minimized to enhance sweep efficiency. Moreover, polymer inclusion enables active surface transformations and catalyses the formation of O/W emulsions [64–67]. Available literature in this field of research indicates successful polymer flooding trails associated with the laboratory, pilot plant and field application-scale systems associated to mature oil fields [18,21,35,68,69]. A review article reported 72 polymer flooding field cases that were undertaken since 1964 (66 onshore and 6 offshore projects) [70]. Among these, while major projects have been undertaken and implemented in the

USA (64%), few studies have been conducted in Canada and China (8%) and Germany (5.6%). Among 72 investigations, 6 cases indicated discouraging outcomes. This is due to identified non-optimal characteristics of the reservoir system, namely too low average permeability (~7 mD), low injected polymer concentration (213 ppm), higher viscosity reduction, high resistance factor, high permeability contrast and higher polymer retention. Also, the investigations primarily targeted the application of 92% HPAM (partially hydrolysed polyacrylamide) and 8% biopolymer. Considering laboratory, pilot plant and field application-scale systems, the authors have reviewed the relevance of polymer flooding operations for the recovery of heavy crude oil since 1960 [35]. Polymer flooding enabled an increase in oil recovery from 2.2% to 44% of OOIP for laboratory-scale systems. However, the same cannot be achieved for the field application system due to many technical issues. Based on the analytical data and influence of parameters, the authors developed generalized rules of thumb for the relevance or irrelevance of polymer flooding to improvise oil recovery from heavy crude oil reservoirs. The generalized rules of thumb according to the authors' analysis refers to the following requirements for the application of polymer flooding: (a) reservoir depth to be <5250 ft, (b) reservoir temperature to be <149°F, (c) reservoir permeability to be >1000 mD, (d) reservoir rock porosity to be >21%, (e) oil gravity to be >11° API, (f) oil viscosity to be <5400 cP, (g) oil saturation to be >50% and (h) formation water to have salinity <46,000 ppm.

The performance of polymer flooding is significantly influenced by the parameters of crude oil and reservoir system. In a few studies, it has been analysed that a salinity and temperature variation from moderate to extreme reservoir conditions can strongly influence the oil recovery characteristics of the polymer flooding system [71–74]. This is possibly due to viscosity reduction of the aqueous system that was accomplished by *in-situ* chemical degradation. Other investigations showed that polymer degradation occurs due to the hydrolysis of amide groups with ions prevalent in the formation waters to generate carboxylic acids in the polymer flooding system [75,76]. Thereby, the aqueous media viscosity gets reduced along with the simultaneous formation of precipitates. Furthermore, the thermal stability of emulsions verily controls the oil recovery factors. At a higher temperature, the de-stability of the emulsions occurs as they undergo coalescence of oil droplets and therefore reduce the oil recovery factor [77,78].

Combined chemical flooding – Flooding studies conducted with a combination of alkali, surfactant and polymer systems indicate better performance (in terms of residual oil recovery) due to the synergy of these constituents in improvising desired characteristics such as improved displacement efficiency, favourable mobility and lower interfacial tension [21,33,79–85].

Sandpack and micro-model flooding experiments were performed to determine residual oil recovery of alkali and alkali–surfactant combination flooding processes [81]. The authors found that alkali reduced IFT to 10^{-2} mN/m; however, for the alkali–surfactant system, the IFT reduced to a very low value of 10^{-4} mN/m. For the 1 wt% NaOH alkali solution case, sandpack flooding indicated the highest recovery of 19.96%, which was higher than that achieved with the combination system of alkali and surfactant (18.63%). The micro-model experiments conveyed that the alkali solution penetrates into the oil phase and facilitates the generation of W/O emulsions to enhance sweep efficiency. Compared to the alkali flooding case, the sweep efficiency reduction was significant for the alkali–surfactant system because of O/W emulsions as confirmed by microscopic flooding tests. Similarly, the synergy of alkali (NaOH)–polymer (HPAM) interaction to enhance crude oil recovery was analysed [21]. Sandpack flooding data with only polymer solution conveyed enhanced tertiary oil recovery (from 4.55% to 16.09%) with increasing HPAM concentration (250–2000 ppm). The polymer solution viscosity verily and greatly contributed to the oil recovery factor. However, for the case of only polymer-containing solution, oil recovery could not be enhanced during flooding due to the viscous fingering effect. This viscous fingering was encountered due to a significant viscosity difference between polymer solution (30 mPa.s) and that of the heavy oil (3950 mPa.s). Moreover, as compared to only polymer flooding, the alkali only flooding provided better oil recovery (30.31% at 1 wt% NaOH). This was due to the reason that alkali enables better penetration of the aqueous solution into the oil phase generating W/O emulsion. This emulsion diverts the aqueous solution to unswept regions covering more reservoir area and ultimately enhances the sweep efficiency. However, alkali only flooding was not significant to maximize the performance of oil recovery as it depends on the flow of W/O emulsion. Therefore, it was due to the combined effect of alkali–polymer flooding that the highest oil recovery of 43.38% was achieved at 1 wt% NaOH and 1000 ppm HPAM formulation. In another similar study, the authors tried to evaluate the synergy of combined chemicals by performing surfactant and surfactant–polymer flooding. Surfactant flooding was successful in recovering 20% OOIP whereas with surfactant–polymer flooding, the recovery factor was further enhanced by 2.78% OOIP for the same system. The formation of oil bank by surfactant and improved sweep efficiency with polymer together maximizes the oil recovery factor [82]. Finally, the synergy of organic alkali–surfactant–polymer (ASP) flooding in Shengli oilfield (China) was characterized with a higher content of divalent ions (Ca^{2+} and Mg^{2+}) [33]. A greater synergy of surfactant (Shengli petroleum sulfonate), organic alkali (ethanolamine) and Shengli crude oil facilitated a significant reduction in the IFT value to about 10^{-3} mN/m. Flooding experiments indicated oil recovery of only 4% for surfactant, 10.7% using polymer, 13.7% using a surfactant–polymer combination and 21.7% (maximum oil recovery of 21.7%) using the ASP flooding process. An enhancement in organic alkali concentration facilitated a further increase in oil recovery to reach the highest value of 23.4%. The microscopic displacement tests conducted converge upon the involved mechanisms for various combinations of chemicals. The authors inferred that W/O emulsion is formed due to the reaction of alkali with the acidic constituents of the crude oil. The emulsion diverts the flooding fluid to the unswept regions and promotes sweep efficiency. The *in-situ* surfactant and induced surfactant together contribute to the reduction of IFT to reach very low values. Thereby, crude oil dispersion is significantly facilitated to enable the generation of O/W emulsions and thereby enables increased displacement efficiency. Without any precipitation of salt, the organic ASP flooding enabled the realization of higher oil recovery. The combined influence of enhanced sweep and displacement efficiencies has been indicated to the most appropriate mechanism associated with the higher oil recovery during organic ASP flooding. Hence, the ASP flooding has been proven to be effective to efficiently recover crude oil from heavy oil reservoirs [86]. The incremental production of oil recovered through ASP flooding has been suggested to be cost effective [80].

Chemical-nanofluid EOR: On a laboratory scale, nanoparticles in chemical-nanofluid flooding systems have been evaluated to function well for crude oil systems being operated with higher combinations of temperature and salinity. Thereby, enhanced residual oil recovery has been reported in comparison with chemical flooding systems [37,38,77]. Researchers have addressed the efficacy of silica nanoparticles (~15 nm) assisted polymer flooding for enhanced oil recovery application [38]. The experimental investigations targeted on the stability, rheology, IFT reduction and flooding methodologies to evaluate cumulative oil recovery. The silica nanoparticle content was varied from 0.5 to 2 wt% for the evaluation of nanoparticle stability in a polymer solution. Initially, through visual observation, the authors investigated on the sedimentation behaviour. They concluded that nanoparticles were stable in polymer solution for a month when the concentration of nanoparticles was around 0.5 wt%. However, the stability time was reduced to 27, 22 and 19 days for 1, 1.5 and 2 wt% cases, respectively. Nanoparticles in polymer solution underwent agglomeration and formed clusters. The average size of these clusters increased with increasing concentration of the nanoparticles and thus enhances the sedimentation phenomena. The zeta potential of the nanoparticle–polymer systems was above –30 mV, which confirmed the system stability. The viscosity of chemical solutions enhanced about three times with the inclusion of nanoparticles in the polymer containing media. Additionally, the effect of temperature on the viscosity variation was not significant in the nanoparticle-containing polymer solution system. The IFT reduction behaviour was observed for the nanoparticle–polymer solutions and the system was monitored for 27 days. The lowest IFT value was detected at 1 wt% SiO_2 nanoparticles, and the IFT reduction continued even at elevated temperature thus indicating the potential of nanoparticles at a higher temperature. The cumulative oil recovery with polymer was 58.17%, and this increased to 62.22% for a 1.0 wt% SiO_2 nanoparticle–polymer containing system. However, at a higher temperature of 90°C, the oil recovery reduced negligibly from 62.22%

to 61.67%. The same system was successful in enhancing the oil recovery to 70.45% with the inclusion of surfactant (surfactant–polymer–nanoparticle system). In summary, the oil recovery determined from nanofluid flooding experiments confirmed the efficacy of nanoparticles for EOR applications including reservoirs at high temperature. Moreover, the relative permeability curve revealed that the system wettability can be potentially altered from intermediate wet state to water-wet state. Several other nanoparticles like TiO_2, Al_2O_3 and SiO_2 have been targeted for EOR application on limestone samples at operating temperatures of 26°C, 40°C, 50°C, and 60°C [87]. Compared to brine, the solution containing alumina, titanium and silica nanoparticle showed a lower IFT and it was reduced by 33%, 37% and 42%, respectively. At 26°C, the displacement test indicated an oil recovery of 51.9%, 50.8% and 48.7% for alumina, titanium and silica nanoparticle-containing solutions, respectively. With an increase in temperature, the oil recovery factor increased and maximum recovery of the crude oil was achieved at 60°C. For alumina, titanium and silica nanoparticle solutions, the crude oil recovery was evaluated as 65.7%, 61.9% and 57.7%, respectively. The nanofluids (nanoparticle-containing aqueous media) behaved as Newtonian fluids in the entire temperature range and were comparatively more viscous (1.28–1.65 cP) in comparison to that of the brine (0.94 cP). The contact angle was lowest for SiO_2, and this value reduced from 26° to 18° for an elevation in system temperature from 26°C to 60°C. Similarly, for TiO_2 and Al_2O_3 cases, the contact angle reduced from 57° to 46° and from 71° to 61°, respectively, with a similar increase in temperature. The study was confined to nanoparticles in an aqueous system and did not target investigations with polymer–nanoparticle systems.

1.7 Complexities and Literature Lacuna for Chemical and Chemical-Nanofluid EOR

To date, several successful chemical flooding investigations have targeted the residual oil recovery characteristics. Despite indicating the potential of alternate mechanisms or their combination for various crude oil-reservoir systems, the literature findings cannot be generalized to address trends apparent for any crude oil-reservoir system. Hence, chemical flooding characteristics always are to be investigated for each individual case, due to the very fact that the EOR characteristics and pertinent mechanisms are invariably dependent upon the combinatorial characteristics of crude oil, rock, temperature and formation water systems.

The available state-of-the-art elaborates upon the relationships between crude oil acid value and residual oil recovery. However, the crude oil naphthenic acid characteristics that vary for each reservoir were not considered in terms of its efficacy for *in-situ* soap formation. Furthermore, stepwise neutralization and saponification of acid groups with alkali were not evaluated which are well known to provide useful insights with respect to simultaneous IFT reduction and emulsification aided with alkali penetration into the oil phase. Such studies can provide a deeper understanding of the associated complexities of the alkali flooding system. Considering the complexities involved in the IFT reduction and emulsification mechanism during chemical flooding, a critical observation of the available literature for higher oil recovery is inconclusive. Advanced studies with surfactant systems declared ultra-low IFT [88]. However, a generalization of such trends cannot be considered with the fact that an alkali–surfactant combination can also reduce IFT to an ultra-low value [16]. In a typical flooding system involving both alkali and surfactant, bettering oil recovery is dependent upon the characteristic properties of both the reservoir and crude oil system. These include the dominance of either or more of the following phenomena: IFT reduction, emulsification, wettability alteration, naphthenic acid content and its role in saponification and neutralization. Also, in the confined field of on-field application of surfactant flooding systems, the available prior art is limited for carbonate reservoirs [89].

The adsorption of surfactants on reservoir rock systems detrimentally influences their technical and economic efficacy. This is due to the very reason that enhanced surfactant adsorption translates into greater surfactant loss, insignificant IFT reduction and emulsification. Hence, it is very important to evaluate the extent of surfactant adsorption on reservoir rock systems. The available prior-art in this area of research involved many assumptions that do not suffice to real-world scenarios. These include consideration of synthetic cores instead of real cores, investigations at room but not elevated temperatures, utilization of aqueous but not simulated reservoir water system, consideration of rock systems with fixed characteristics associated with mineral content and surface area. Furthermore, the surfactant adsorption characteristics need to target desired attributes such as reduced adsorption capacity, greater stability of surfactants and ageing characteristics at elevated temperatures [59].

Recent investigations targeted nanoparticle-assisted chemical flooding and declared that the technology has a promising impact to improvise upon the residual oil recovery factor [38,90–93]. As compared to conventional chemical flooding, nanoparticles have the unique ability to improve system rheology, reduce IFT and alter wettability to facilitate the achievement of additional recovery of crude oil during flooding studies. However, the crude oil–chemical–nanoparticle system synergy has not been extensively evaluated from the perspectives of emulsification stability like detailing of droplet coalescence with nanoparticles [94]. Accordingly, there is a definite need to critically examine the tradeoffs of chemical flooding systems with nanoparticles. In this regard, it can be conceptually analysed that while higher nanoparticle content may enhance emulsification and IFT reduction characteristics, they are very likely to clog the porous structure of the reservoir system and thereby hamper residual recovery. Consequently, optimal nanoparticle content needs to be achieved in laboratory and field-scale investigations.

In summary, a critical insight into the available prior research indicates that both chemical flooding and nanoparticle-integrated chemical flooding systems have received little attention in terms of their efficacy associated with combinatorial performance characteristics such as reduction of IFT, alteration in wettability, stability of emulsions, rheological analysis and tertiary oil recovery. Considering these limitations, chemical and chemical-nanoparticle systems have been discussed in a detailed and systematic format throughout this book. The critical findings have been addressed to reduce the knowledge gap and highlight the research gap in the relevant field.

References

1. U.S. Energy Information Administration, Short-Term Energy Outlook, May 2016 and January 2017.
2. J. J. Sheng, *Modern Chemical Enhanced Oil Recovery Theory and Practice*. Gulf Professional Publishing, Elsevier, Houston, TX, 2011.
3. Indian Petroleum and Gas Statistics FY13, Published on December 2013.
4. A. O. Al-Amodi, U. A. Al-Mubaiyedh, A. S. Sultan, M. S. Kamal, and I. A. Hussein, Novel fluorinated surfactants for enhanced oil recovery in carbonate reservoirs, *Canadian Journal of Chemical Engineering*, vol. 94, pp. 454–460, 2016.
5. S. Kumar and A. Mandal, Studies on interfacial behavior and wettability change phenomena by ionic and nonionic surfactants in presence of alkalis and salt for enhanced oil recovery, *Applied Surface Science*, vol. 372, pp. 42–51, 2016.
6. S. Iglauer, Y. Wu, P. Shuler, Y. Tang, and W. A. Goddard III, Alkyl polyglycoside surfactant–alcohol cosolvent formulations for improved oil recovery, *Colloids and Surfaces A: Physicochemical and Engineering Aspects*, vol. 339, pp. 48–59, 2009.
7. J. Raicar and R. M. Procter, Economic considerations and potential of heavy oil supply from Lloydminster – Alberta, Canada, in Meyer, R. F., Wynn, J. C., Olson, J. C. (Eds.), *The Second UNITAR International Conference on Heavy Crude and Tar Sands, Caracas*. McGraw-Hill, New York, pp. 212–219, February 7–17, 1984.
8. A. Hemmati-Sarapardeh, H. Hashemi Kiasari, N. Alizadeh, S. Mighani, and A. Kamari, *Application of fast-SAGD in naturally fractured heavy oil reservoirs: A case study*, Presented at the 18th Middle East Oil & Gas Show and Conference, Manama, Bahrain, March 10–13, 2013.
9. A. Hemmati-Sarapardeh, M. Khishvand, A. Naseri, and A. H. Mohammadi, Toward reservoir oil viscosity correlation, *Journal of Chemical Engineering Science*, vol. 90, pp. 53–68, 2013.

10. J. J. Taber, F. D. Martin, and R. S. Seright, EOR screening criteria revisited – part 1: Introduction to screening criteria and enhanced recovery field projects, *SPE Reservoir Engineering*, vol. 12, pp. 189–198, 1997.
11. Sino Australia Oil and Gas Limited, An Introduction to Enhanced Oil Recovery Techniques, 2013.
12. L. Lake, *Enhanced Oil Recovery*. Prentice Hall, Hoboken, NJ, 1989.
13. D. W. Green and G. P. Willhite, *Enhanced Oil Recovery*, vol. 6. SPE Textbook Series. SPE, Richardson, TX, 1998.
14. C. E. Johnson, Status of caustic and emulsion methods, *Journal of Petroleum Technology*, vol. 28, pp. 85–92, 1976.
15. C. E. Cooke, R. E. Williams, and P. A. Kolodzie, Oil recovery by alkaline waterflooding, *Journal of Petroleum Technology*, vol. 26, pp. 1365–1374, 1974.
16. H. A. Nasr-El-Din and K. C. Taylor, Dynamic interfacial tension of crude oil/alkali/surfactant systems, *Colloids and Surfaces*, vol. 66, pp. 23–37, 1992.
17. D. J. Pye, Improved secondary recovery by control of water mobility, *Journal of Petroleum Technology*, vol. 16, pp. 911–916, 1964.
18. H. L. Chang, Polymer flooding technology yesterday, today, and tomorrow, *Journal of Petroleum Technology*, vol. 30, pp. 1–16, 1978.
19. J. J. Sheng, Investigation of alkaline-crude oil reaction, *Petroleum*, vol. 1, pp. 31–39, 2015.
20. C. D. Yuan, W. F. Pu, X. C. Wang, L. Sun, Y. C. Zhang, and S. Cheng, Effects of interfacial tension, emulsification, and surfactant concentration on oil recovery in surfactant flooding process for high temperature and high salinity reservoirs, *Energy and Fuels*, vol. 29, pp. 6165–6176, 2015.
21. H. Pei, G. Zhang, J. Ge, L. Zhang, and H. Wang, Effect of polymer on the interaction of alkali with heavy oil and its use in improving oil recovery, *Colloids and Surfaces A: Physicochemical and Engineering Aspects*, vol. 446, pp. 57–64, 2014.
22. M. Zargartalebi, N. Barati, and R. Kharrat, Influences of hydrophilic and hydrophobic silica nanoparticles on anionic surfactant properties: Interfacial and adsorption behaviors, *Journal of Petroleum Science & Engineering*, vol. 119, pp. 36–43, 2014.
23. A. Bera, K. Ojha, A. Mandal, and T. Kumar, Interfacial tension and phase behavior of surfactant-brine–oil system, *Colloids and Surfaces A: Physicochemical and Engineering Aspects*, vol. 383, pp. 114–119, 2011.
24. A. M. Bellocq, D. Bourbon, B. Lemanceau, and G. Fourche, Thermodynamic, interfacial, and structural properties of polyphasic microemulsion systems, *Journal of Colloid and Interface Science*, vol. 89, pp. 427–440, 1982.
25. J. Ge, A. Feng, G. Zhang, P. Jiang, H. Pei, R. Li, et al., Study of the factors influencing alkaline flooding in heavy-oil reservoirs, *Energy and Fuels*, vol. 26, pp. 2875–2882, 2012.
26. M. Dong, Q. Liu, and A. Li, Displacement mechanisms of enhanced heavy oil recovery by alkaline flooding in a micromodel, *Particuology*, vol. 10, pp. 298–305, 2012.
27. J. Wang, M. Dong, and M. Arhuoma, Experimental and numerical study of improving heavy oil recovery by alkaline flooding in sandpacks, *Journal of Canadian Petroleum Technology*, vol. 49, pp. 51–57, 2010.
28. J. J. Taber, F. D. Martin, and R. S. Seright, EOR screening criteria revisited-part 2: Applications and impact of oil prices, SPE-39234, *in SPE/DOE Improved Oil Recovery Symposium*, 21–24 April, Tulsa, Oklahoma, 1997.

29. H. Sharma, S. Dufour, G. W. P. P. Arachchilage, U. Weerasooriya, G. A. Pope, and K. Mohanty, Alternative alkalis for ASP flooding in anhydrite containing oil reservoirs, *Fuel*, vol. 140, pp. 407–420, 2015.
30. A. A. Olajire, Review of ASP EOR (alkaline surfactant polymer enhanced oil recovery) technology in the petroleum industry: Prospects and challenges, *Energy*, vol. 77, pp. 963–982, 2014.
31. A. Bera, A. Mandal, and B. B. Guha, Synergistic effect of surfactant and salt mixture on interfacial tension reduction between crude oil and water in enhanced oil recovery, *Journal of Chemical & Engineering Data*, vol. 59, pp. 89–96, 2014.
32. S. Zendehboudi, M. A. Ahmadi, A. R. Rajabzadeh, N. Mahinpey, and I. Chatzis, Experimental study on adsorption of a new surfactant onto carbonate reservoir samples-application to EOR, *Canadian Journal of Chemical Engineering*, vol. 91, pp. 1439–1449, 2013.
33. L. Fu, G. Zhang, J. Ge, K. Liao, H. Pei, P. Jiang, et al., Study on organic alkali-surfactant-polymer flooding for enhanced ordinary heavy oil recovery, *Colloids and Surfaces A: Physicochemical and Engineering Aspects*, vol. 508, pp. 230–239, 2016.
34. I. A. Malik, U. A. Al-Mubaiyedh, A. S. Sultan, M. S. Kamal, and I. A. Hussein, Rheological and thermal properties of novel surfactant-polymer systems for EOR applications, *The Canadian Journal of Chemical Engineering*, vol. 94, pp. 1693–1699, 2016.
35. H. Saboorian-Jooybari, M. Dejam, and Z. Chen, Heavy oil polymer flooding from laboratory core floods to pilot tests and field applications: Half-century studies, *Journal of Petroleum Science & Engineering*, vol. 142, pp. 85–100, 2016.
36. S. R. Clark, M. J. Pitts, and S. M. Smith, Design and application of an alkaline-surfactant-polymer recovery system to the west Kiehl field, *SPE Advanced Technology Series*, vol. 1, pp. 172–179, 1993.
37. R. Saha, R. Uppaluri, and P. Tiwari, Silica nanoparticle assisted polymer flooding of heavy crude oil: Emulsification, rheology, and wettability alteration characteristics, *Industrial and Engineering Chemistry Research*, vol. 57, pp. 6364–6376, 2018.
38. T. Sharma, S. Iglauer, and J. S. Sangwai, Silica nanofluids in an oilfield polymer polyacrylamide: Interfacial properties, wettability alteration, and applications for chemical enhanced oil recovery, *Industrial & Engineering Chemistry Research*, vol. 55, pp. 12387–12397, 2016.
39. M. Tang, G. Zhang, J. Ge, P. Jiang, Q. Liu, H. Pei, et al., Investigation into the mechanisms of heavy oil recovery by novel alkaline flooding, *Colloids and Surfaces A: Physicochemical and Engineering Aspects*, vol. 421, pp. 91–100, 2013.
40. M. S. Almalik, A. M. Attia, and L. K. Jang, Effects of alkaline flooding on the recovery of Safaniya crude oil of Saudi Arabia, *Journal of Petroleum Science and Engineering* vol. 17, pp. 367–372, 1997.
41. D. Xie, J. Hou, F. Zhao, A. Doda, and J. Trivedi, The comparison study of IFT and consumption behaviors between organic alkali and inorganic alkali, *Journal of Petroleum Science & Engineering*, vol. 147, pp. 528–535, 2016.
42. H. Y. Jennings, C. E. Johnson, and C. D. McAuliffe, A caustic waterflooding process for heavy oils, *Journal of Petroleum Technology*, vol. 26, pp. 1344–1352, 1974.
43. J. Rudin and D. T. Wasan, Mechanisms for lowering of interfacial tension in alkali/acidic oil systems 2: Theoretical studies, *Colloids and Surfaces*, vol. 68, pp. 81–94, 1992.

44. H. Pei, G. Zhang, J. Ge, L. Jin, and C. Ma, Potential of alkaline flooding to enhance heavy oil recovery through water-in-oil emulsification, *Fuel,* vol. 104, pp. 284–293, 2013.
45. H. Gong, Y. Li, M. Dong, S. Ma, and W. Liu, Effect of wettability alteration on enhanced heavy oil recovery by alkaline flooding, *Colloids and Surfaces A: Physicochemical and Engineering Aspects,* vol. 488, pp. 28–35, 2016.
46. R. Ehrlich and R. J. Wygal, Interrelation of crude oil and rock properties with the recovery of oil by caustic waterflooding, in *SPE-AIME Fourth Symposium on Improved-Oil Recovery,* 22–24 March, Tulsa, 1976, pp. 263–270.
47. G. Jerauld and J. Rathmell, Wettability and relative permeability of prudhoe bay: A case study in mixed-wet reservoirs, *SPE Reservoir Engineering,* vol. 12, pp. 58–65, 1997.
48. N. R. Morrow, Wettability and its effect on oil recovery, *Journal of Petroleum Technology,* vol. 42, pp. 1476–1484, 1990.
49. T. Austad, B. Matre, J. Milter, A. Saevareid, and L. Oyno, Chemical flooding of oil reservoirs 8: Spontaneous oil expulsion from oil- and water-wet low permeable chalk material by imbibition of aqueous surfactant solutions, *Colloids and Surfaces A: Physicochemical and Engineering Aspects,* vol. 137, pp. 117–129, 1998.
50. M. Budhathoki, T.-P. Hsu, P. Lohateeraparp, B. L. Roberts, B.-J. Shiau, and J. H. Harwell, Design of an optimal middle phase microemulsion for ultra high saline brine using hydrophilic lipophilic deviation (HLD) method, *Colloids and Surfaces A: Physicochemical and Engineering Aspects,* vol. 488, pp. 36–45, 2016.
51. Q. Liu, M. Dong, S. Ma, and Y. Tu, Surfactant enhanced alkaline flooding for western canadian heavy oil recovery, *Colloids and Surfaces A: Physicochemical and Engineering Aspects,* vol. 293, pp. 63–71, 2007.
52. L. Chen, G. Zhang, J. Ge, P. Jiang, J. Tang, and Y. Liu, Research of the heavy oil displacement mechanism by using alkaline/surfactant flooding system, *Colloids and Surfaces A: Physicochemical and Engineering Aspects,* vol. 434, pp. 63–71, 2013.
53. S. A. Pursley and H. L. Graham, Borregos field surfactant pilot test, *Journal of Petroleum Technology,* vol. 27, pp. 695–700, 1975.
54. J. M. Maerker and W. W. Gale, Surfactant flood process design for loudon, *SPE Reservoir Engineering,* vol. 7, pp. 36–44, 1992.
55. W. W. Gale and E. I. Sandvik, Tertiary surfactant flooding: Petroleum sulfonate composition-efficacy studies, *Society of Petroleum Engineers Journal,* vol. 13, pp. 191–199, 1973.
56. D. C. Standnes and T. Austad, Wettability alteration in chalk 2: Mechanism for wettability alteration from oil-wet to water-wet using surfactants, *Journal of Petroleum Science & Engineering,* vol. 28, pp. 123–143, 2000.
57. K. Jarrahian, O. Seiedi, M. S. Sheykhan, M. V. Sefti, and S. Ayatollahi, Wettability alteration of carbonate rocks by surfactants: A mechanistic study, *Colloids and Surfaces A: Physicochemical and Engineering Aspects,* vol. 410, pp. 1–10, 2012.
58. R. Saha, R. Uppaluri, and P. Tiwari, Impact of natural surfactant (reetha), polymer (xanthan gum), and silica nanoparticles to enhance heavy crude oil recovery, *Energy and Fuels,* vol. 33, pp. 4225–4236, 2019.
59. M. S. Kamal, A. S. Sultan, and I. A. Hussein, Screening of amphoteric and anionic surfactants for cEOR applications using a novel approach, *Colloids and Surfaces A: Physicochemical and Engineering Aspects,* vol. 476, pp. 17–23, 2015.

60. W. Pu, C. Yuan, W. Hu, T. Tan, J. Hui, S. Zhao, et al., Effects of interfacial tension and emulsification on displacement efficiency in dilute surfactant flooding, *RSC Advances*, vol. 6, pp. 50640–50649, 2016.
61. S. Park, E. S. Lee, and W. R. W. Sulaiman, Adsorption behaviors of surfactants for chemical flooding in enhanced oil recovery, *Journal of Industrial and Engineering Chemistry*, vol. 21, pp. 1239–1245, 2015.
62. T. Amirianshoja, R. Junin, A. K. Idris, and O. Rahmani, A comparative study of surfactant adsorption by clay minerals, *Journal of Petroleum Science & Engineering*, vol. 101, pp. 21–27, 2013.
63. B. B. Sandiford, Laboratory and field studies of water floods using polymer solutions to increase oil recoveries, *Journal of Petroleum Technology*, vol. 16, pp. 917–922, 1964.
64. E. Akiyama, A. Kashimoto, K. Fukuda, H. Hotta, T. Suzuki, and T. Kitsuki, Thickening properties and emulsification mechanisms of new derivatives of polysaccharides in aqueous solution, *Journal of Colloid and Interface Science*, vol. 282, pp. 448–457, 2005.
65. E. Akiyama, T. Yamamoto, Y. Yago, H. Hotta, T. Ihara, and T. Kitsuki, Thickening properties and emulsification mechanisms of new derivatives of polysaccharide in aqueous solution 2: The effect of the substitution ratio of hydrophobic/hydrophilic moieties, *Journal of Colloid and Interface Science*, vol. 311, pp. 438–446, 2007.
66. R. S. Seright, T. G. Fan, K. Wavrik, H. Wan, N. Gaillard, and C. Favero, Rheology of a new sulfonic associative polymer in porous media, *SPE Reservoir Evaluation & Engineering*, vol. 14, pp. 726–734, 2011.
67. Y. Lu, W. Kang, J. Jiang, J. Chen, D. Xu, P. Zhang, et al., Study on the stabilization mechanism of crude oil emulsion with an amphiphilic polymer using the b-cyclodextrin inclusion method, *RSC Advances*, vol. 7, pp. 8156–8166, 2017.
68. R. L. Jewett and G. F. Schurz, Polymer flooding-A current appraisal, *Journal of Petroleum Technology*, vol. 22, pp. 675–684, 1970.
69. W. Demin, Z. Zhang, L. Chun, J. Cheng, X. Du, and Q. Li, A pilot for polymer flooding of saertu formation S II 10–16 in the north of Daqing oil field, in *SPE Asia Pacific Oil and Gas Conference*, 28–31 October, Adelaide, Australia, 1996, pp. 431–441.
70. D. C. Standnes and I. Skjevrak, Literature review of implemented polymer field projects, *Journal of Petroleum Science & Engineering*, vol. 122, pp. 761–775, 2014.
71. W. M. Leung and D. E. Axelson, Thermal degradation of polyacrylamide and poly (acrylamide-coacrylate), *Journal of Polymer Science Part A*, vol. 25, pp. 1852–1864, 1987.
72. M. H. Yang, Rheological behavior of polyacrylamide solution, *Journal of Polymer Engineering*, vol. 19, pp. 371–381, 1999.
73. H. Kheradmand, J. Francois, and V. Plazanet, Hydrolysis of polyacrylamide and acrylic acid-acrylamide copolymers at neutral pH and high temperature, *Polymer*, vol. 29, pp. 860–870, 1988.
74. Q. Chen, Y. Wang, Z. Lu, and Y. Feng, Thermoviscosifying polymer used for enhanced oil recovery: Rheological behaviors and core flooding test, *Polymer Bulletin*, vol. 70, pp. 391–401, 2013.
75. G. Muller, Thermal stability of high molecular weight polyacrylamide aqueous solution, *Polymer Bulletin*, vol. 5, pp. 31–37, 1981.

76. B. F. Abu-Sharkh, G. O. Yahaya, S. A. Ali, E. Z. Hamad, and I. M. Abu-Reesh, Viscosity behavior and surface and interfacial activities of hydrophobically modified water-soluble acrylamide/N-phenyl acrylamide block copolymers, *Journal of Applied Polymer Science*, vol. 89, pp. 2290–2300, 2002.
77. T. Sharma, G. S. Kumar, B. H. Chon, and J. S. Sangwai, Thermal stability of oil-in-water pickering emulsion in the presence of nanoparticle, surfactant, and polymer, *Journal of Industrial and Engineering Chemistry*, vol. 22, pp. 324–334, 2015.
78. B. P. Binks and A. Rocher, Effects of temperature on water-in-oil emulsions stabilised solely by wax microparticles, *Journal of Colloid and Interface Science*, vol. 335, pp. 94–104, 2009.
79. Y. Wu, M. Dong, and S. Ezeddin, Study of alkaline/polymer flooding for heavy-oil recovery using channeled sandpacks, in *Canadian Unconventional Resources and International Petroleum Conference*, Calgary, 2010.
80. M. J. Pitts, P. Dowling, K. Wyatt, H. Surkalo, and C. Adams, Alkaline-surfactante-polymer flood of the tanner field, in *SPE/DOE Symposium on Improved Oil Recovery*, 22–26 April, Tulsa, Oklahoma, 2006, pp. 1–5.
81. H. Pei, G. Zhang, J. Ge, M. Tang, and Y. Zheng, Comparative effectiveness of alkaline flooding and alkaline–surfactant flooding for improved heavy-oil recovery, *Energy and Fuels*, vol. 26, pp. 2911–2919, 2012.
82. A. Samanta, K. Ojha, A. Sarkar, and A. Mandal, Surfactant and surfactant-polymer flooding for enhanced oil recovery, *Advances in Petroleum Exploration and Development*, vol. 2, pp. 13–18, 2011.
83. M. Pratap and M. S. Gauma, Field implementation of alkaline-surfactant-polymer (ASP) flooding: A maiden effort in India, SPE-88455, in *SPE Asia Pacific Oil and Gas Conference and Exhibition*, Perth, Austrlia, 18–20 October, 2004, pp. 1–5.
84. H. F. Li, G. Z. Liao, P. H. Han, Z. Y. Yang, X. L. Wu, G. Y. Chen, et al., Alkaline/surfactant/polymer (ASP) commercial flooding test in the central Xing2 area of daqing oilfield, SPE-84896, in *International Improved Oil Recovery Conference in Asia Pacific*, 20–21 October, Kuala Lumpur, Malaysia, 2003.
85. Z. J. Yang, Y. Zhu, X. Y. Ma, X. H. Li, J. Fu, P. Jiang, et al., ASP pilot test in Guodao field, *Journal of Jianghan Petroleum Institute*, vol. 24, pp. 62–64, 2002.
86. H. S. Hadi, M. Degam, and Z. X. Chen, Heavy oil polymer flooding from laboratory core floods to pilot tests and field applications: Half-century studies, *Journal of Petroleum Science and Engineering*, vol. 142, pp. 85–100, 2016.
87. A. E. Bayat, R. Junin, A. Samsuri, A. Piroozian, and M. Hokmabadi, Impact of metal oxide nanoparticles on enhanced oil recovery from limestone media at several temperatures, *Energy Fuels* vol. 28, pp. 6255–6266, 2014.
88. D. M. Wang, C. D. Liu, W. X. Wu, and G. Wang, Development of an ultra-low interfacial tension surfactant in a system with no-alkali for chemical flooding, in *SPE/DOE Improved Oil Recovery Symposium*, Tulsa, Oklahoma, 2008, pp. 1–9.
89. J. J. Sheng, Review of surfactant enhanced oil recovery in carbonate reservoirs, *Advances in Petroleum Exploration and Development*, vol. 6, pp. 1–10, 2013.
90. L. Hendraningrat, S. Li, and O. Torsaeter, A coreflood investigation of nanofluid enhanced oil recovery, *Journal of Petroleum Science and Engineering*, vol. 111, pp. 128–138, 2013.
91. H. Yousefvand and A. Jafari, Enhanced oil recovery using polymer/nanosilica, *Procedia Materials Science*, vol. 11, pp. 565–570, 2015.

92. A. Maghzi, R. Kharrat, A. Mohebbi, and M. H. Ghazanfari, The impact of silica nanoparticles on the performance of polymer solution in presence of salts in polymer flooding for heavy oil recovery, *Fuel*, vol. 123, pp. 123–132, 2014.
93. N. Kumar, T. Gaur, and A. Mandal, Characterization of SPN pickering emulsions for application in enhanced oil recovery, *Journal of Industrial and Engineering Chemistry* vol. 54, pp. 304–315, 2017.
94. Y. Sun, D. Yang, L. Shi, H. Wu, Y. Cao, Y. He, et al., Properties of nano-fluids and their applications in enhanced oil recovery: A comprehensive review, *Energy and Fuels*, vol. 34, pp. 1202–1218, 2020.

2

Alkali Flooding – Mechanisms Investigation

2.1 Introduction to Alkali Flooding

Alkali flooding is a fascinating area of research that shows promising results towards higher oil recovery in both laboratory and pilot/oil field scales. A significant attraction to execute alkali flooding was recognized in different countries like China, USA, Saudi Arabia, Canada and so on [1–5]. The most commonly used alkalis that are deployed for flooding are NaOH [6], Na_2CO_3 [7], NaCl [5], (9) KOH [4], $NaB(OH)_4$ [8] and Na_3PO_4 [4], and their success depends on the nature of crude oil and reservoir properties. Researchers have suggested different mechanisms that are responsible for higher oil recovery during alkali flooding. The mechanisms include the reduction in interfacial tension (IFT) between oil and the aqueous alkaline phase [6,9], emulsification and entrapment [6], emulsification and entrainment [10], emulsification and coalescence [11] and wettability alteration (water-wet to oil-wet [5,12] or oil-wet to water-wet [13–15], whichever is favourable).

The investigation on alkali flooding to recover residual oil was initiated in 1927 by Atkinson [16]. Alkali responds to the natural acids that exist in crude oil to form in-situ soap, which reduces the IFT and thus enhances the overall recovery of crude oil [6,9,17]. An ultra-low IFT value was approached when both ionized and unionized acids are adsorbed at the interface resulting in mixed micelles [18]. The oil–alkali chemistry monitors IFT reduction, which favours the fractional flow and hence improves the oil recovery factor [19]. Therefore to understand the oil–alkali chemistry, the role of acid value of oil recovery during alkali flooding was investigated. It was reported that crude oil with an acid value of ≥1.5 was favourable in recovering additional oil by the mechanism of emulsion bank formation [5]. Moreover, it was also reported that the residual oil recovery for heavy oil reservoirs could be enhanced from 12.4% to 20.4% as the acid value increases from 1.84 to 4.66 mg of KOH/g of the sample, respectively [20]. Thus it was observed that IFT reduction in addition to emulsification is responsible for improved oil recovery [2,4,21].

Alkali while flowing in the reservoir has the ability to penetrate inside the crude oil medium, which can result in the formation of an emulsion bank. This emulsion increases the aqueous phase flow resistance, which routes the aqueous phase to the unswept region of the reservoirs improving the sweep. Therefore, during alkali flooding, the synergy of IFT reduction and emulsification mechanisms together enhances the residual oil recovery [3,21–23]. Additionally, the recovery of residual oil was also found to be dependent on the wettability of the reservoir. A study reported that during alkali flooding, the alteration in wettability of the system from water-wet to the favourable oil-wet can result in better recovery of crude oil [12]. A detailed summary of the screening of alkali flooding is provided in Table 2.1 [24].

2.2 IFT between Crude Oil and Alkaline Solution

The reduction in IFT is one of the important and essential conditions for successful alkali flooding. The concept of alkali flooding was introduced in 1927 by Atkinson where they studied the importance of IFT reduction phenomena, which leads to higher recovery of crude oil [16]. The pertinent mechanism of alkali flooding is depicted in Figure 2.1. As shown, alkali reacts with the naphthenic acid group prevalent in crude oil and facilitates the formation of an in-situ soap at the interface which decreases IFT [6,9,16,17,19]. The reaction steps to understand the alkali–crude oil chemistry while forming the in-situ soap are studied and explained in detail by Sheng [33].

In our study, light-moderate crude oil and formation water collected from an Indian oil field (Assam, India) were used to study the IFT. The density, API gravity, acid value, viscosity and surface tension of the crude oil were found to be 892.4 kg/m^3, 27.06°, 2.94 mg of KOH/g sample, 33.4 mPa·s at 25°C and 27.3 mN/m respectively. The reservoir formation water, which was used to prepare alkali solutions, was collected from the Assam reservoir. The compositional analysis of the formation water consists of Na^+ = 1308 ppm, K^+ = 65 ppm, Ca^{2+} = 18 ppm, Mg^{2+} = 12 ppm and Cl^- = 876 ppm. The initial IFT between crude oil and reservoir water was found to be 16.3 mN/m. The reduction in IFT with variation in alkali (NaOH) concentrations from 0.2 to 1 wt% is shown in Figure 2.2. Figure 2.2a represents the compared dynamic IFT behaviour at 0.8 wt% NaOH from different studies [21,34]. The dynamic IFT of the system initially decreases, and then it starts to increase with time before it finally flattens off to reach equilibrium [34]. This pattern is usually encountered due to the adsorption–desorption behaviour of active surfactant species at the interface [21,35,36]. The initial decrease in IFT was because of the higher adsorption rate of the formed in-situ soap at the interface. Thus, a significant growth of in-situ surfactants is achieved at the interface, which assists in reducing the IFT. As the build-up of in-situ surfactant increases with time, a higher concentration gradient is developed at the interface, which then enhances the rate of desorption. This desorption reduces the quantity of active soap at the interface and thus enhances the IFT. Equilibrium IFT was

TABLE 2.1

Screening Criteria Summary for Alkali Flooding [24]

Lithology	Depth (ft)	Clay	Permeability of Formation (mD)	Formation Water Salinity (TDS, ppm)	Divalent (ppm)	Reservoir Temperature (°C)	Oil Viscosity (CP)	API Gravity	Acid No.	Oil Saturation (S_o)	Aquifer	Gas Cap	References
Sandstone			≥20			≤93.3	≤200	≤35	≥0.2		None to weak	None to weak	[25]
Sandstone	<4921		>30	Low	Low	40–60	2-150	>27.5	1.5-4	>(0.3-0.5)	Weak	Weak	[26]
Sandstone			>20	<100,000		<93.3	<90	<30					[27]
Sandstone preferred	<9000		>20			<93.3	90	<30	Organic acid	>S_{or}			[28]
Sandstone preferred	<9000		>10			<93.3	<35	>20	Organic acid	>0.35			[29,30][a]
Sandstone		Low	>50	50,000	1000	<70	<150	<35		0.35	None	None	[31][b]
	500–9000		>100	<200 k if T<60°C, <50 k if T>60°C		<93.3	<35	>20		>0.45			[32][b]
Majority sandstone	2650	Low	240	24,313	145	45.2	17	22.3	0.82	0.52	Generally none	Generally none	Alkaline projects
Sandstone	NC	Low	>10	<50,000	<100	<93.3	<150	NC	Organic acid	>0.35	Weak	Weak	Proposed for alkaline

Source: (Reprinted from J. J. Sheng, "Status of Alkaline Flooding Technology," Journal of Petroleum Engineering & Technology, vol. 5, pp. 44–50, 2015.)

[a] For alkaline, SP and ASP.

[b] For ASP, NC = not critical.

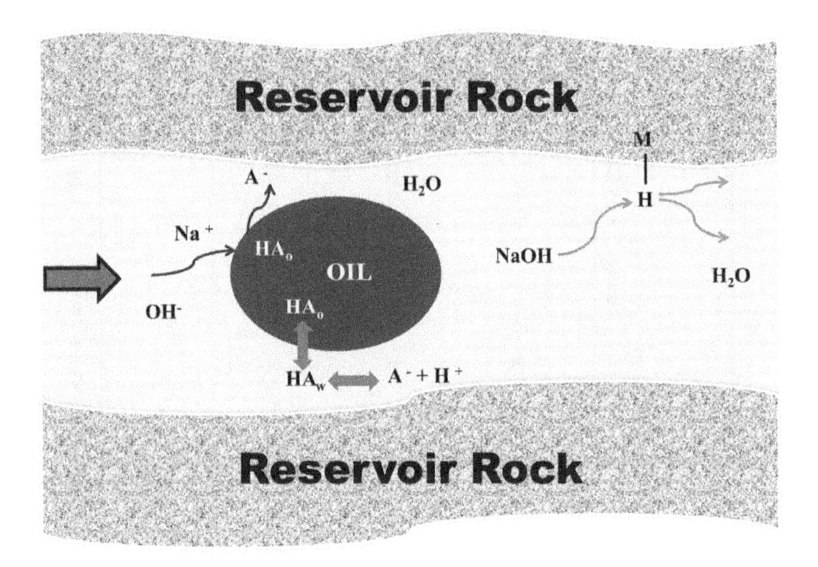

FIGURE 2.1
Schematic representation of the crude oil–alkali reaction mechanism inside a porous media of reservoir rock [19].

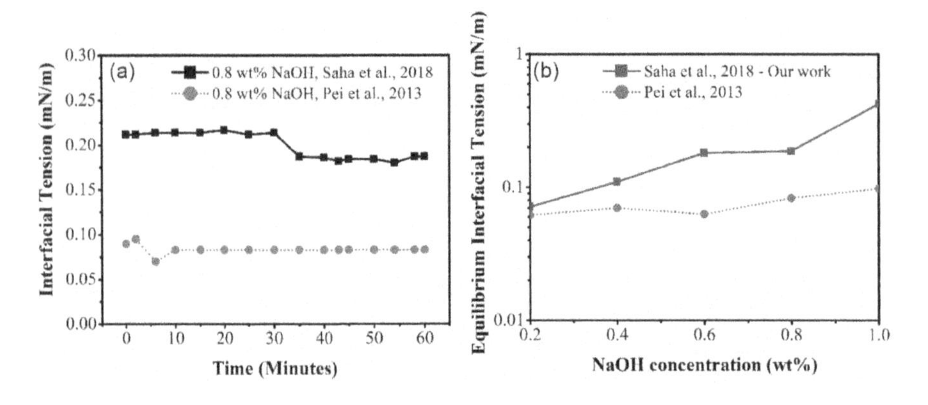

FIGURE 2.2
IFT behaviour using alkali (NaOH) (a) dynamic IFT at 0.8 wt% and (b) equilibrium IFT at different NaOH concentrations.

reached when the rate of adsorption and desorption are equal and hence no further changes in the IFT can be observed thereafter.

Figure 2.2b shows the variation in equilibrium IFT within the range of 0.06–0.42 mN/m as the concentration of NaOH changes from 0.2 to 1.0 wt%. The IFT at 0.1 wt% alkali could not be measured as in such case no oil drops were formed. The lowest IFT value of 7.1×10^{-2} mN/m was detected at a lower concentration of 0.2 wt% NaOH. As the concentration of alkali increases, the IFT starts to enhance and continues till 1 wt% NaOH. This IFT behaviour depends on the pH of the system. Initially, the pH of the system is less because of lower alkali concentration and this leads to a greater amount of unionized acids followed by a reduction in the critical micelle concentration (CMC) of ionized acids. Hence, this results in a minimum interfacial coverage and a greater IFT value. However, at a higher concentration of alkali solution, the interfacial coverage is enhanced, which reduces the IFT to optimum pH. Beyond optimum pH, the IFT again increases, which is due to higher or complete ionization of the acid group that results in a minimum quantity of unionized acid [18,21,37].

2.3 Alkali Flooding in Sandpack

As majority of the reservoirs on which EOR has been conducted are of sandstone type, sandpack has been used to execute flooding experiments. The elemental compositional analysis of the sand particles showed that oxygen (53.9%) and silica (45%) are the major elements with minor traces of Al (0.6%), Fe (0.4%) and Na (0.1%). The effectiveness of alkali (NaOH) flooding to recover residual oil on a laboratory scale for Assam crude oil is estimated by performing sandpack flooding runs as depicted in Table 2.2. The permeability of the sandpack ranges from 1.9 to 2.1 D, and the saturation of oil was around 79%–84%. The oil recovery by water flooding was in the range of 32%–35% original oil in place (OOIP).

2.3.1 Alkali Concentration on Oil Recovery

Alkali flooding experiments were conducted by varying the alkali concentration in the range of 0.2 to 1 wt% NaOH (Table 2.2). A significant amount of residual oil was recovered as the concentration of NaOH enhances. The

TABLE 2.2

Summary of Oil Recovery Percentage by Alkali (NaOH) Flooding Using Sandpack

Serial Number	Water Flooding Recovery (OOIP %)	NaOH Concentration (wt%)	Alkali Recovery (OOIP %)	Cumulative Oil Recovery (OOIP %)
1	33.4	0.2	11.3	44.7
2	34.2	0.4	14.3	48.5
3	35.0	0.6	21.9	56.9
4	33.5	0.8	24.6	58.1
5	32.9	1	25.5	58.4

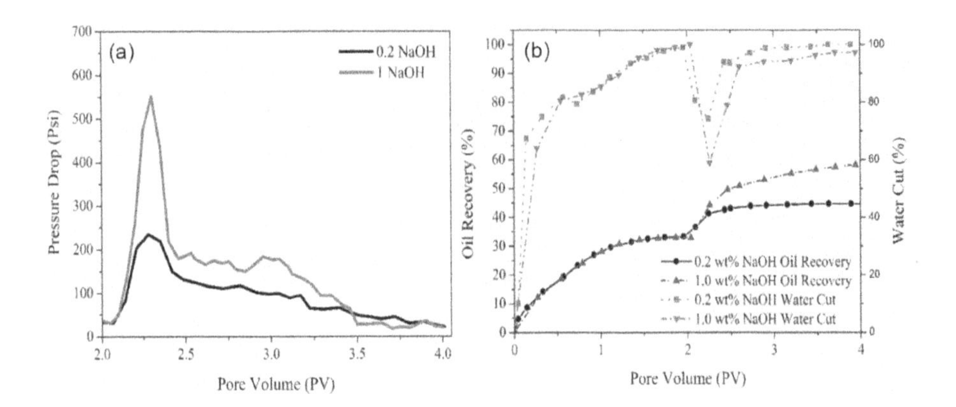

FIGURE 2.3
Curve obtained during alkali flooding experiments at 0.2 and 1 wt% NaOH (a) pressure drop curve and (b) oil recovery and water cut curve.

oil recovery approaches greater than double as the amount of NaOH boosts from 0.2 to 0.8 wt%. This enhancement in recovery factor with alkali concentration is because of the pH effect and piercing of NaOH in the crude oil medium to form emulsion of water in oil (W/O) type (Figure 2.4). A stagnant point in the oil recovery was observed beyond 0.8 wt% because of the emulsion quality, which controls the sweep efficiency. Hence, a highest oil recovery factor of 25.5% oil was recovered at 1 wt% NaOH. The maximum oil recovered can be further explained by the pressure drop and water cut curve as provided in Figure 2.3a and b, respectively. The higher pressure drop indicates piercing of alkali solution in the crude oil medium, which results in the formation of high-viscous W/O emulsion. This emulsion thereby introduces favourable mobility ratio, which improves the overall sweep efficiency and ultimately results in higher oil recovery [21]. The enhanced oil recovery trend with alkali concentration has been validated by other authors as well [2,7,21,38]. However, flooding runs beyond 1 wt% NaOH are not encouraged as in such conditions, the emulsion quality may not be effective to improve the sweep efficiency.

2.3.2 Extent of Emulsification and Size Distribution of Droplets

The extent of emulsification was observed by examining the emulsion produced during alkali flooding experiments. Figure 2.4 illustrates a schematic diagram indicating the development of W/O (water in oil) emulsion in which water is dispersed in the form of droplets with oil being the continuous medium. The droplet size distributions of the emulsions were examined by analysing the microscopy images obtained at various NaOH concentrations [34]. The average droplet diameter of the emulsion was less (2.85 μm) at 0.2 wt% NaOH, which successively increased to 5.63 μm as the alkali concentration enhances to 1 wt% [34]. The droplet size distribution was in the

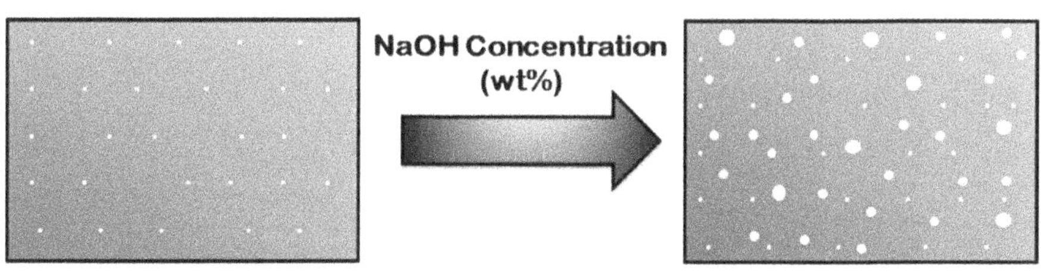

NaOH Concentration	Droplet Size Distribution Range (μm)	Average Droplet Size (μm)
0.2	0 - 7	2.85
0.4	0 - 11	4.03
0.6	1 - 12	4.45
0.8	1 - 15	4.95
1.0	0 - 15	5.63

FIGURE 2.4
A schematic diagram representing the type of emulsion formed, the distribution of the droplet sizes and the average droplet size formed during alkali flooding experiments at different NaOH concentrations.

range of 0–7 μm (small) for a lower concentration of 0.2 wt% NaOH, which then increased to 0–15 μm (large) as the concentration of NaOH approached 1 wt% NaOH (Figure 2.4). Therefore, the droplet data clearly demonstrate the emulsification extent with the alkali concentration. This emulsification extent enhances the emulsion quality by increasing the average droplet size of the emulsion that enables the discharge of trapped crude oil from the reservoir and thereby results in improved oil recovery [3,21–23].

Researchers have worked intensively on the impact of viscosity of the W/O emulsion for its flow through the reservoir pore spaces. In their study, they observed that emulsion viscosity is directly proportional to the emulsion quality [3,39]. A better emulsion quality is required, which has the potential to divert the aqueous phase covering more area of the reservoirs. This reduces the viscous fingering effect and thereby enhances the sweep efficiency [23,40,41].

2.3.3 Slug Size on Residual Oil Recovery

Considering the economy of the process, it is advisable to inject a small volume of alkali slug, which has the potential to maximize the oil recovery factor. Therefore, to achieve the optimization of flooding conditions, variation in slug size at the optimum alkali concentration (1 wt% NaOH) was investigated and the corresponding residual oil recoveries were reported as depicted in Figure 2.5. Initially at a low slug volume of 0.25 PV, the residual oil recovery was 20.7% IOIP (initial oil in place), which then enhances to 25.5% IOIP as

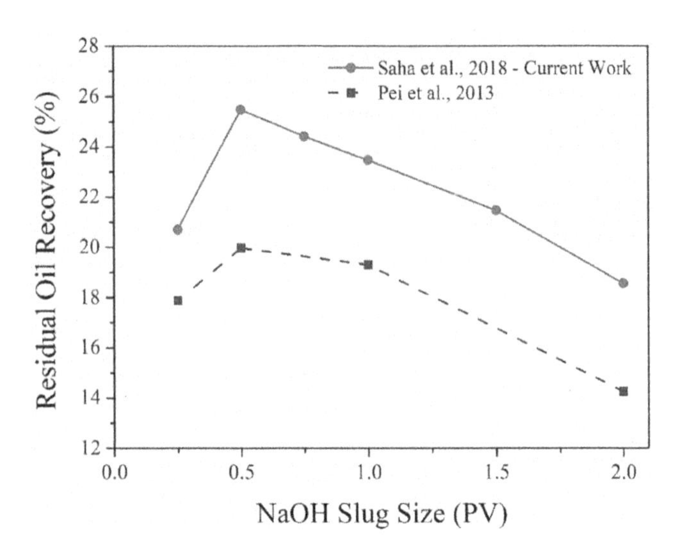

FIGURE 2.5
Influence of alkali pore volume on the recovery of crude oil during NaOH flooding.

the slug volume increases to 0.5 PV. A maximum oil recovery was obtained during such scenario and any further increase in slug volume reduces the recovery factor significantly. A lowest reduced residual oil recovery of 18.5% IOIP was obtained as the slug volume reaches 2 PV [34]. Similar behaviour on residual oil recovery was reported by other studies while performing NaOH flooding as shown in Figure 2.5 [21]. Initially, as the slug volume increases from 0.25 to 0.5 PV, the emulsion formed (W/O type) possesses the characteristics to reduce viscous fingering by blocking water channelling. This phenomenon increases the sweep efficiency by initiating the aqueous medium to come in contact with the uncovered region of the crude oil reservoir, thus enhancing oil recovery. However, the formed emulsion also possesses negative effect that can severely affect the oil recovery factor. At a higher slug volume of above 0.5 PV, greater amounts of W/O emulsion are formed, which are highly viscous, and this creates poor mobility declining the oil recovery factor [21]. Hence, formulation of optimum slug was accomplished in order to achieve the maximum recovery of crude oil.

2.3.4 Injection Pattern on Oil Recovery

The cyclic and continuous injection pattern has been investigated by performing flooding experiments at 1 wt% NaOH. The flooding experiments include two shots of 0.25 PV (total 0.5 PV) for the cyclic injection system and one shot of 0.5 PV for the continuous injection system as represented in Figure 2.6. The cumulative oil recovered with cyclic injection was 53.9% and for continuous injection, the recovery was 58.4% IOIP. The cyclic injected resulted in lower recovery, which is because of the emulsion break down

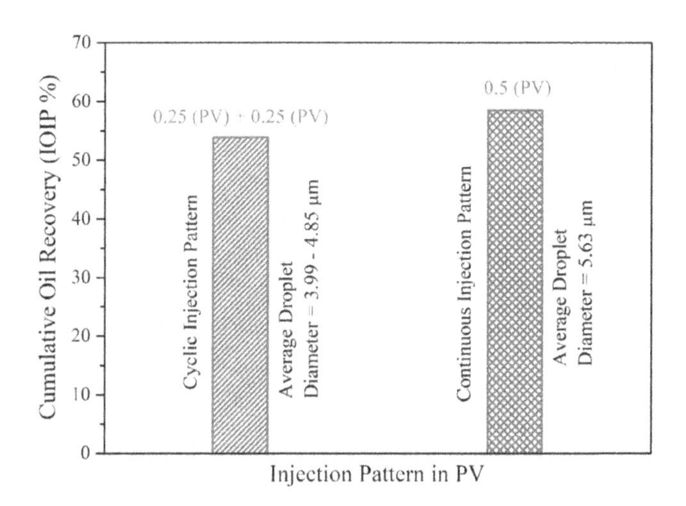

FIGURE 2.6
Influence of the alkali injection pattern on the recovery of crude oil.

that drastically impacted the plugging effect. The breakdown was predicted by analysing the emulsions formed during the cyclic injection pattern. The average diameters of the droplets differ in between 3.99 and 4.85 μm, which clearly demonstrated the non-uniformity of the emulsion droplets [34]. The cyclic injection further introduces a dilution effect of the alkali slug, which hampers the formation of W/O emulsion [21]. The continuous injection pattern is therefore advisable as in such a scenario, emulsion breakdown was not encountered and the average diameter of the droplets was uniform with 5.63 μm size as illustrated in Figures 2.4 and 2.6.

2.3.5 Alkali Injection Rate for Oil Recovery

The rate at which the alkali slug is injected inside the core samples during flooding experiments decides the amount of residual oil recovery. Initially, at a low injection rate, the emulsion formed is severely affected as the driving force is low [21]. As the injection rate increases, an optimum rate is achieved during which the emulsion formed is sufficient enough to block the water channelling and increases the flow resistance. This reduces viscous fingering, increases sweep efficiency and eventually results in a higher oil recovery [5,7,21]. At a higher injection rate, the contact time and area reduce because of which the alkali solution could not react properly with the acid group of the oil. This behaviour affects the emulsion formation capacity and results in emulsion breakdown causing early water breakthrough. Therefore, the combination of all these factors results in poor sweep efficiency lowering the oil recovery factor [7,21]. Researchers have also inspected the connection between injection rate and pressure drop across the sandpack during flooding experiments. They observed that at a higher injection rate, the pressure drop is about three times higher than the normal injection rate. However, practically it is not possible to enhance the oil recovery at a higher injection rate though it is pressure sensitive [7].

2.4 Neutralization and Saponification of Crude Oil

The detection of a carboxylic acid group in crude oil is already reported in the FTIR spectrum of our earlier publication [34]. This carboxylic acid interacts with alkali to produce in-situ soap, which reduces the IFT and enhances the recovery of crude oil. Therefore, the role of the acid group present in the oil can be visualized by investigating its interaction with the alkali solutions. The quantification of neutralization of the acid group and estimation of the saponification formation is essential to understand the oil–water (alkaline) chemistry.

The naphthenic acid of the crude oil is 0.35 mg of KOH/g, which corresponds to around 12% of the total acid value. The neutralization and saponification extent of the oil varied with variation in the concentration of NaOH. At a lower concentration of 0.2 wt% NaOH, the amount of acid which takes part in saponification is only 14.6%. The acid reacts and undergoes further neutralization/saponification up to 34.5% as the concentration of alkali reaches 1 wt% NaOH. However, around 65% of the acid groups are still active to undergo neutralization. The mechanism of oil–alkali chemistry is depicted and explained in Figure 2.1 [19]. The in-situ soap formed during the reaction was detected by analyzing the aqueous effluent collected during alkali flooding experiments through an FTIR spectrum [34].

2.5 Wettability Alteration of Reservoir Rock

The alteration in wettability of reservoir rock can be estimated by determining the angle of contact between the rock surface (saturated with oil) and the drop of alkali solution as shown in Table 2.3. The variation in contact angle at different alkali concentrations was detected up to 15 minutes and beyond that, no further or negligible changes in contact angle were observed. Hence, the final contact angle for different systems was considered at 15 minutes.

The transformation in the contact angle from 109° to 75.2° using reservoir formation water (0% NaOH) indicates the wettability of the reservoir as intermediate wet. However, for solutions at different alkali concentrations, the final contact angle obtained was 53.9° at 0.2 wt%, 41.5° at 0.4 wt%, 38.9° at 0.6 wt%, 37.3° at 0.8 wt% and 36.8° at 1.0 wt%. As the concentration of alkali increases, the drop spreads out on the surface resulting in wettability

TABLE 2.3

Contact Angle Value with Time at Different Concentrations of NaOH

Alkali Concentration (wt%)	Contact Angle (°)					Final Contact Angle (°)
	$T = 0$ minutes	$T = 5$ minutes	$T = 10$ minutes	$T = 15$ minutes	$T = 20$ minutes	
Formation water	109.3	89.7	81.2	75.2	74.8	Formation water, 75.2°
0.2	94.7	76.5	61.4	53.9	52.8	0.2 wt% NaOH, 53.9°
0.4	90.4	72.4	60.9	41.5	41.2	0.4 wt% NaOH, 41.5°
0.6	83.2	61.3	44.8	38.9	38.2	0.6 wt% NaOH, 38.9°
0.8	89.8	51.4	44.5	37.3	36.6	0.8 wt% NaOH, 37.3°
1.0	84.8	56.2	45.2	36.8	36.4	1.0 wt% NaOH, 36.8°

alteration of the system from initial intermediate wet to final water wet [42]. The enhanced alkali concentration facilitates the formation of in-situ soap, which then diffuses in the aqueous media and improves the hydrophilicity. Furthermore, the adsorption of this in-situ soap on the solid mineral surface changes the contact angle [12]. Thus, the overall effect causes the trapped oil to flow through the porous media and increases the recovery of crude oil.

2.6 Overall Factors Deciding Oil Recovery

The potential of different properties like IFT, emulsion, alteration in wettability and saponification has been summarized to estimate the recovery factor of crude oil. Initially, at a lower concentration of 0.2 wt% NaOH, the lowest IFT of 7.1×10^{-2} mN/m was detected, which corresponds to emulsion formation of 2.85 μm droplet diameter followed by 54° contact angle and 14.61% saponification. The oil recovery under such a scenario was lowest and it was estimated to be about 11.3% IOIP. However, as the concentration of alkali enhances to 1 wt% NaOH, the IFT reaches 0.425 mN/m with 5.63 μm droplet size emulsions followed by 37° contact angle and 34% saponification. The recovery of crude oil was maximized and it was detected to be around 25.5% IOIP. The output results specify the importance of IFT reduction for oil recovery but attaining the lowest IFT does not maximize the recovery of crude oil. Moreover, the increase in alkali concentration improves the fostering of saponification, which simultaneously enhances the emulsion quality. Additionally, the drastic reduction in contact angle signifies the alteration in wettability from intermediate to the desired water wet. Thus, it was due to the simultaneous effect of all the above-mentioned properties, which leads towards the highest recovery of crude oil and not the individual influence of IFT reduction [12,21,40,43]. Hence, the importance of acid content is significant as it greatly monitors the saponification, which directly decides the emulsion quality.

The available literature reported several successful alkali flooding for crude oil of light, light-moderate and heavy categories. The light crude oil from the Louisiana oil field (USA) was successful to recover 28% additional oil. Similar enhanced oil recoveries of 26% and 21% were obtained for light-moderate crude oil in the field of Texas-Frlo and Louisiana field respectively [43]. In the case of a heavy oil field, around 15%–30% additional oil was encountered in the fields of China and Canada [2,3,21,44]. However, from the available literature, we cannot evaluate a generalized thumb rule which can predict the success of alkali flooding. Hence, individual case studies and the organisation of these data from various oil fields can be clubbed together to develop different correlations that can be deployed for alkali flooding depending on the reservoir properties.

2.7 Conclusions

The study focused on the impact of different mechanisms such as reduction in IFT, emulsion quality, alteration in wettability and saponification towards the quantification of crude oil recovery by alkali flooding. The acid value of the crude oil (Assam, India) was quantified and its roles in fostering the mechanisms responsible for oil recovery were examined. The data revealed simultaneous participation of different mechanisms for the current crude oil system through which additional or tertiary oil recovery of around 25.5% was encountered. The reduction in IFT assists the recovery factor of crude oil but the lowest IFT alone could not trigger the recovery. The lowest IFT was not successful in enhancing the emulsion quality and improving the saponification and neither could alter the wettability of the system. Therefore, to enhance all these factors, the alkali concentration was enhanced and thus the maximized recovery factor was obtained. Finally, the flooding parameters were explored to optimize the system and the optimum value was discovered to be 0.5 PV slug size with continuous injection of the alkali slug. The variation in the value of contact angle showed the visibility of alteration of the reservoir from intermediate to the desirable water wet.

References

1. C. E. Johnson, Status of caustic and emulsion methods, *Journal of Petroleum Technology*, vol. 28, pp. 85–92, 1976.
2. M. Tang, G. Zhang, J. Ge, P. Jiang, Q. Liu, H. Pei, et al., Investigation into the mechanisms of heavy oil recovery by novel alkaline flooding, *Colloids and Surfaces A: Physicochemical and Engineering Aspects*, vol. 421, pp. 91–100, 2013.
3. J. Wang, M. Dong, and M. Arhuoma, Experimental and numerical study of improving heavy oil recovery by alkaline flooding in sandpacks, *Journal of Canadian Petroleum Technology*, vol. 49, pp. 51–57, 2010.
4. M. S. Almalik, A. M. Attia, and L. K. Jang, Effects of alkaline flooding on the recovery of Safaniya crude oil of Saudi Arabia, *Journal of Petroleum Science & Engineering* vol. 17, pp. 367–372, 1997.
5. C. E. Cooke, R. E. Williams, and P. A. Kolodzie, Oil recovery by alkaline waterflooding, *Journal of Petroleum Technology*, vol. 26, pp. 1365–1374, 1974.
6. H. Y. Jennings, C. E. Johnson, and C. D. McAuliffe, A caustic waterflooding process for heavy oils, *Journal of Petroleum Technology*, vol. 26, pp. 1344–1352, 1974.
7. M. Dong, S. Ma, and Q. Liu, Enhanced heavy oil recovery through interfacial instability: A study of chemical flooding for Brintnell heavy oil, *Fuel*, vol. 88, pp. 1049–1056, 2009.
8. F. Chen, H. Jiang, X. Bai, and W. Zheng, Evaluation the performance of sodium metaborate as a novel alkali in alkali/surfactant/polymer floodin, *Journal of Industrial and Engineering Chemistry*, vol. 19, pp. 450–457, 2013.
9. H. Y. Jennings, A study of caustic solution-crude oil interfacial tensions, *Society of Petroleum Engineers Journal*, vol. 15, pp. 197–202, 1975.
10. P. Subkow, Process for the removal of bitumen from bituminous deposits, US Patent No. 2,288,857, 1942.
11. T. P. Castor, W. H. Somerton, and J. F. Kelly, Recovery mechanisms of alkaline flooding, in Shah, D. O. (Ed.), *Surface Phenomena in Enhanced Oil Recovery*. Plenum Press, New York, pp. 249–291, 1981.
12. H. Gong, Y. Li, M. Dong, S. Ma, and W. Liu, Effect of wettability alteration on enhanced heavy oil recovery by alkaline flooding, *Colloids and Surfaces A: Physicochemical and Engineering Aspects*, vol. 488, pp. 28–35, 2016.
13. N. Mungan, Certain wettability effects in laboratory waterfloods, *Journal of Petroleum Technology*, vol. 18, pp. 247–252, 1966.
14. N. Mungan, Interfacial effects in immiscible liquid-liquid displacement in porous medium, *Society of Petroleum Engineers Journal*, vol. 6, pp. 247–253, 1966.

15. O. R. Wagner and R. O. Leach, Improving oil displacement by wettability adjustment, *Society of Petroleum Engineers,* vol. 216, pp. 65–72, 1959.
16. H. Atkinson, Recovery of Petroleum from Oil-Bearing Sands, US1651311 1927.
17. P. G. Nutting, Chemical problems in the water driving of petroleum from oil sands, *Industrial and Engineering Chemistry,* vol. 10, pp. 1035–1036, 1925.
18. J. Rudin and D. T. Wasan, Mechanisms for lowering of interfacial tension in alkali/acidic oil systems 2: Theoretical studies *Colloids and Surfaces,* vol. 68, pp. 81–94, 1992.
19. E. F. DeZabala, J. M. Vislocky, E. Rubin, and C. J. Radke, A chemical theory for linear alkaline flooding, *Society of Petroleum Engineers,* vol. 22, pp. 245–258, 1982.
20. J. Ge, A. Feng, G. Zhang, P. Jiang, H. Pei, R. Li, et al., Study of the factors influencing alkaline flooding in heavy-oil reservoirs, *Energy and Fuels,* vol. 26, pp. 2875–2882, 2012.
21. H. Pei, G. Zhang, J. Ge, L. Jin, and C. Ma, Potential of alkaline flooding to enhance heavy oil recovery through water-in-oil emulsification, *Fuel,* vol. 104, pp. 284–293, 2013.
22. M. Dong, Q. Liu, and A. Li, Displacement mechanisms of enhanced heavy oil recovery by alkaline flooding in a micromodel, *Particuology,* vol. 10, pp. 298–305, 2012.
23. H. Pei, G. Zhang, J. Ge, L. Jin, and X. Liu, Analysis of microscopic displacement mechanisms of alkaline flooding for enhanced heavy-oil recovery, *Energy and Fuels,* vol. 25, pp. 4423–4429, 2011.
24. J. J. Sheng, Status of alkaline flooding technology, *Journal of Petroleum Engineering & Technology,* vol. 5, pp. 44–50, 2015.
25. Enhanced oil recovery, *National Petroleum Congress,* Technical Report, USA, NPC, 1976.
26. A. N. Carcoana, Enhanced oil recovery in Rumania, SPE-10699, in *SPE Enhanced Oil Recovery Symposium,* 4–7 April, Tulsa, Oklahoma, 1982.
27. Enhanced oil recovery, *National Petroleum Congress,* Technical Report, USA, NPC, 1984.
28. G. O. Goodlett, M. M. Honarpour, F. T. Chung, and P. S. Sarathi, The role of screening and laboratory flow studies in EOR process evaluation, SPE-15172, in *SPE Rocky Mountain Regional Meeting,* 19–21 May, Billings, Montana, 1986.
29. J. J. Taber, F. D. Martin, and R. S. Seright, EOR screening criteria revisited - part1: Introduction to screening criteria and enhanced recovery field projects, *SPE Reservoir Engineering,* vol. 12, pp. 189–198, 1997.
30. J. J. Taber, F. D. Martin, and R. S. Seright, EOR screening criteria revisited-part 2: Applications and impact of oil prices, SPE-39234, in *SPE/DOE Improved Oil Recovely Symposium,* 21–24 April, Tulsa, Oklahoma, 1997.
31. M. A. Al-Bahar, R. Merrill, W. Peake, and M. Jumaa, Evaluation of IOR potential within Kuwait, SPE-88716, in *Abu Dhabi International Conference and Exhibition,* 10–13 October, Abu Dhabi, United Arab Emirates, 2004.
32. J. L. Dickson, A. Leahy-Dios, and P. L. Wylie, Development of improved hydrocarbon recovery screening methodologies, SPE-129768, in *SPE Improved Oil Recovery Symposium,* 24–28 April, Tulsa, Oklahoma, 2010.
33. J. J. Sheng, *Modern Chemical Enhanced Oil Recovery Theory and Practice.* Gulf Professional Publishing, Elsevier, Houston, TX, 2011.

34. R. Saha, R. V. S. Uppaluri, and P. Tiwari, Influence of emulsification, interfacial tension, wettability alteration and saponification on residual oil recovery by alkali flooding, *Journal of Industrial and Engineering Chemistry*, vol. 59, pp. 286–296, 2018.
35. H. A. Nasr-El-Din and K. C. Taylor, Dynamic interfacial tension of crude oil/alkali/surfactant systems, *Colloids and Surfaces*, vol. 66 pp. 23–37, 1992.
36. R. P. Borwankar and D. T. Wasan, Dynamic interfacial tensions in acidic crude oil/caustic systems part I: A chemical diffusion-kinetic model, *AIChE Journal*, vol. 32, pp. 455–466, 1986.
37. J. Rudin and D. T. Wasan, Mechanisms for lowering of interfacial tension in alkali acidic oil system 1: Experimental studies, *Colloids and Surfaces*, vol. 68, pp. 67–79, 1992.
38. A. Samanta, K. Ojha, and A. Mandal, Interactions between acidic crude oil and alkali and their effects on enhanced oil recovery, *Energy and Fuels* vol. 25, pp. 1642–1649, 2011.
39. M. Arhuoma, M. Dong, D. Yang, and R. Idem, Determination of water-in-oil emulsion viscosity in porous media, *Industrial & Engineering Chemistry Research*, vol. 48, pp. 7092–7102, 2009.
40. H. Pei, G. Zhang, J. Ge, M. Tang, and Y. Zheng, Comparative effectiveness of alkaline flooding and alkaline–surfactant flooding for improved heavy-oil recovery, *Energy and Fuels*, vol. 26, pp. 2911–2919, 2012.
41. H. Pei, G. Zhang, J. Ge, M. Ma, L. Zhang, and Y. Liu, Improvement of sweep efficiency by alkaline flooding for heavy oil reservoirs, *Journal of Dispersion Science and Technology*, vol. 34, pp. 1548–1556, 2013.
42. S. Kumar and A. Mandal, Studies on interfacial behavior and wettability change phenomena by ionic and nonionic surfactants in presence of alkalis and salt for enhanced oil recovery, *Applied Surface Science*, vol. 372, pp. 42–51, 2016.
43. R. Ehrlich and R. J. Wygal, Interrelation of crude oil and rock properties with the recovery of oil by caustic waterflooding, *in SPE-AIME Fourth Symposium on Improved- Oil Recovery*, 22–24 March, Tulsa, 1976, pp. 263–270.
44. H. Pei, G. Zhang, J. Ge, L. Zhang, and H. Wang, Effect of polymer on the interaction of alkali with heavy oil and its use in improving oil recovery, *Colloids and Surfaces A: Physicochemical and Engineering Aspects*, vol. 446, pp. 57–64, 2014.

3

Alkali and Surfactant Flooding

3.1 Introduction to Alkali-Surfactant Flooding

An alkali solution when injected into reservoirs generates in-situ soap after reacting with the naphthenic acid of crude oil. The formation of in-situ soap in the aqueous–crude oil interface reduces the interfacial tension (IFT) and forms microemulsion, which thereby controls the recovery mechanism. However, the recovery of residual crude oil can be maximized only when there is a formation of an optimum emulsion (type III) at an optimum salinity during which lowest ultra-low IFT values are encountered. The microemulsion formed while executing alkali flooding may not be at the optimum salinity and hence can severely affect the overall oil recovery factor [1]. Therefore, these deviations can be neglected by adding a synthetic surfactant to the alkali solution. Moreover, the adsorption sites of the solid rock surfaces are saturated by alkali, which reduces the loss of surfactants significantly. The phase behaviour study using alkali ($Na_2O.SiO_2$), synthetic surfactant (NEODOL 25-3S) and Gulf Coast crude oil at different salinities is demonstrated using an activity map [1,2].

The synergy of alkali–surfactant flooding has been proven effective for recovering heavy residual crude oil from oil reservoirs [3–5]. Additionally, alkali reduces the active adsorption sites of the rock minerals, which thereby reduces surfactant loss occurred by adsorption. The available literature on alkali–surfactant flooding concluded that IFT reduction can increase the residual oil recovery and others stated that emulsification of crude oil decides the recovery factor by controlling the sweep efficiency [5–8]. Researchers while investing the alkali–surfactant synergistic effect observed that the IFT of the system can be reduced to 10^{-4}mN/m and during such condition, a highest oil recovery of 19.4% was achieved. In the same system when IFT was reduced further, residual oil recovery decreased because of the emulsion quality that severely affected the sweep efficiency. Thus, no proper correlation between IFT reduction and oil recovery was observed [5]. A similar alkali–surfactant synergy effect for heavy crude oil was reported by researchers, which showed an improved residual oil recovery at a lower IFT of 10^{-2}mN/m. To further understand the complex mechanisms, a comparison study with alkali only and alkali–surfactant combination was performed to check the difference in residual oil recovery. The oil recovery achieved during the sandpack flooding was the highest for alkali-only flooding though ultra-low IFT was detected with the alkali–surfactant system. The W/O emulsion produced during alkali flooding improves the sweep efficiency by reducing viscous fingering, whereas with the alkali–surfactant solution, the W/O emulsion so formed inhibited the flow resulting in poor mobility [6,8]. Hence, we can say that the alkali–surfactant flooding thus results in higher oil recovery [9], but it also has a detrimental effect on oil recovery [6]. Therefore, proper selection of alkali and surfactant considering reservoir salinity and temperature is crucial to enhance the overall cumulative heavy oil recovery in an economical way.

Currently, researchers are exploring carbonate reservoirs as the production of crude oil from sandstone reservoirs is becoming difficult [10–12]. A large portion of crude oil is reserved in carbonate reservoirs and so far, limited work has been carried out for such a category of oil reservoirs. A carbonate reservoir is extremely heterogeneous and possesses low porosity when compared to sandstone porosity. A limited investigation has been carried out on alkali–surfactant flooding for carbonate reservoirs, which revealed the accountability of IFT reduction, emulsification and wettability alteration mechanisms for improved oil recovery [13,14]. A recent article on alkali (Na_2CO_3)–surfactant (CTAB, SDS and TX-100) flooding for carbonate reservoirs detected ultra-low IFT and alteration in wettability, which resulted in improved oil recovery. In the study, wettability alteration played a dominant role in oil recovery when compared to IFT reduction [13]. Thus, considering the available literature, this chapter will discuss and visualize the complexities involved in the alkali and surfactant mixture flooding in carbonate reservoirs for heavy crude oil.

A systematic methodology was adopted to observe the interaction of different alkalis, surfactants and their combinations with heavy crude oil for carbonate reservoirs and will be demonstrated. Initially, the discussion on alkali selection based on the lowest IFT values at different salinities and temperatures will be executed. In a similar manner, the selection of the desired surfactant from different ionic and non-ionic surfactants based on IFT reduction values and emulsification extent will be described. The synergy interaction of the desired alkali and surfactant mixture will then be observed and the corresponding residual oil recovery will be estimated. Finally, the role of surfactant mixture in oil recovery will also be focused on. In all the cases, the active mechanisms like ultra-low IFT, emulsification, displacement efficiency and alteration in reservoir wettability that are responsible for higher oil recovery will be examined in detail.

3.2 Selection of Alkali Based on IFT

3.2.1 IFT between Crude Oil and Different Alkalis

The reaction of alkali with crude oil to reduce IFT is discussed in this section. Figure 3.1 depicts the general behaviour of IFT reduction between crude oil and most commonly used aqueous alkali solutions (NaOH and Na_2CO_3) at different concentrations. Heavy crude oil from Assam oil field (India) with a density of 926.6 kg/m^3, an API gravity of 21.2°, an acid number of 2.72 mg KOH/g sample and viscosity of 20.1 mPa·s at 30°C was considered for our study [15]. The results obtained by us were compared with published data for validation. Moreover, a detailed discussion on those results will assist towards in-depth understanding of the state-of-the-art of the chosen system [6,16–19].

The initial IFT or surface tension for most of the crude oil usually falls in the range of 20–35 mN/m. This IFT then reduces to various ranges depending on the reaction between the crude oil and alkali solutions. Figure 3.1a depicts the reduction in IFT behaviour between sodium hydroxide and crude oil from various oil fields. The minimum equilibrium IFT for Indian crude oil was detected to be 0.53 mN/m, whereas for other crude oil, it reduces to almost 10^{-2} mN/m [6,15,18,19]. Similar IFT reduction was observed with sodium carbonate where the lowest IFT of 1.16 mN/m was detected for Indian crude oil as depicted in Figure 3.1b [15]. However, for other crude oil, minimum IFT was in the range of 8–10^{-2} mN/m [16–18]. Therefore, comparing the data obtained in our study and from the selected literature, we observed that NaOH was better in terms of IFT reduction as compared to Na_2CO_3. Similar behaviour has been observed by other authors which proved better performance of NaOH with respect to Na_2CO_3 [18].

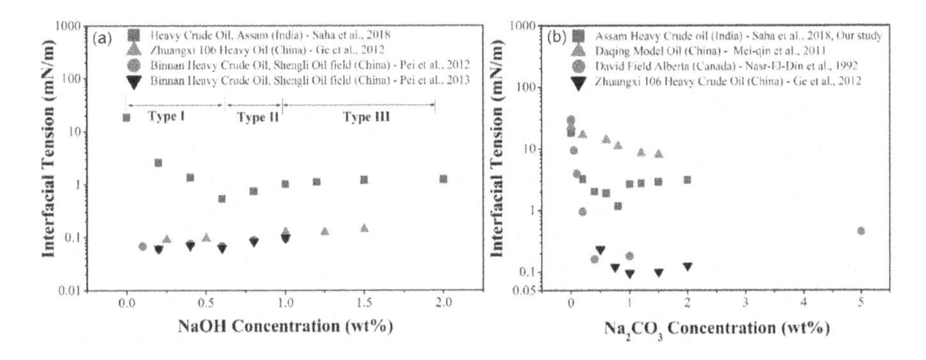

FIGURE 3.1
Reduction in interfacial tension (IFT) between crude oil (oil from a different field) and alkali solutions. (a) Equilibrium IFT at different NaOH concentrations and (b) equilibrium IFT at different Na_2CO_3 concentrations.

The pattern for IFT reduction can be explained by three regions as illustrated in Figure 3.1a. The IFT initially reduces with an increase in alkali concentration (type I). After a certain concentration, the lowest IFT value will be detected (type II) and further increment in alkali concentration thereafter enhances the IFT (type III) [3,16,20]. The reduction in IFT at higher alkali concentrations was due to the pH effect of the solutions. The increase in pH with alkali concentration favours ionization of the acid group present in the crude oil, which reduces the IFT. However, an optimum pH exists during which the lowest IFT can be obtained because of the adsorption of both ionized and un-ionized acids at the interface, with the concentration of un-ionized acid being too low to influence the critical micelle concentration (CMC). A further increase in the concentration of alkali beyond optimum minimizes the amount of un-ionized ions at the interface and therefore the value of IFT increases. Finally, a saturation point in the IFT value was achieved at a higher alkali concentration because of the ionization of all non-ionized acids [19,21,22].

All the IFT data reported were at equilibrium time but the dynamic IFT behaviour usually continued for a period of a maximum of 60 minutes depending on the alkali–crude oil interaction [6,15,18,19]. The variation profile in the dynamic IFT can be explained by the formation of in-situ soap at the oil–water interface and the rate at which the adsorption and desorption occur. Initially, the IFT reduces because of the accumulation of active species at the interface causing a lower desorption rate. However, as time increases, the concentration gradient of the in-situ soap develops because of which the desorption rate enhances. This phenomenon reduces the amount of active species at the oil–water interface, which results in higher IFT [16]. Finally, when the rate of adsorption and desorption reaches equilibrium, the IFT reaches a stagnant point and no further changes can be observed.

3.2.2 Temperature and Salinity Effect on Alkali–Crude IFT

The effects of temperature and salinity on alkali–crude IFT were analysed using sodium hydroxide. Sodium hydroxide was selected for this task as it showed better IFT reduction with Indian crude oil (Assam) as compared to sodium carbonate (see Figure 3.1a and b). As depicted in Figure 3.2, the IFT of the system (NaOH – Indian Crude Oil) enhances from 0.53 to 2.2 mN/m as temperature increases from 30°C to 80°C, respectively. Initially, at room temperature (30°C), the IFT was lowest because of the formation of a stable rigid film at the oil–water interface encountered by the accumulation of cations. However, as temperature increases, the stable rigid film undergoes destabilization and hence enhances the IFT [23,24]. Similar behaviour of IFT increment was observed by other authors as depicted in Figure 3.2 [18]. Moreover, the IFT deviation varied almost negligibly (0.58–0.24 mN/m) with variation in salinity from 0 to 20 wt% [15]. This insignificant reduction in IFT

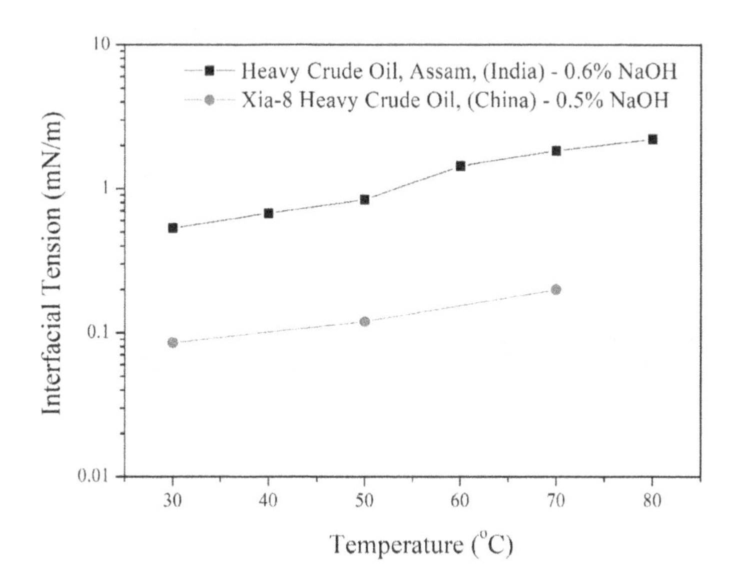

FIGURE 3.2
Interfacial tension between alkali (NaOH) and crude oil at different temperatures.

with salinity could be due to ionic solution, which pushes the in-situ soap at the oil–water interface and enhances the amount of active species at the interface [25,26].

3.3 Selection of Surfactants Based on IFT

3.3.1 IFT between Crude Oil and Different Surfactants

Synthetic and natural surfactants were examined and screened with respect to IFT reduction for heavy crude oil (Assam, India). A total of eight different synthetic surfactants (one cationic, two anionic and five non-ionic) and one natural surfactant (Reetha) were examined and their corresponding reduced equilibrium IFT values are reported in Figure 3.3. Another synthetic surfactant Titriplex III was examined but its IFT could not be recorded as it did not form any droplets while performing IFT experiments.

The selected surfactants were successful in reducing the IFT and the equilibrium IFT reported was in the range of 0.06–7 mN/m. All the synthetic surfactants showed promising IFT reduction up to 0.05 wt%, and on further increasing the surfactant concentration, the IFT remains more or less the same. The initial pattern of IFT reduction with surfactants was because of the enhancement in the adsorption capacity of the surfactant molecules at the oil–water interface. A further rise in the surfactant concentration

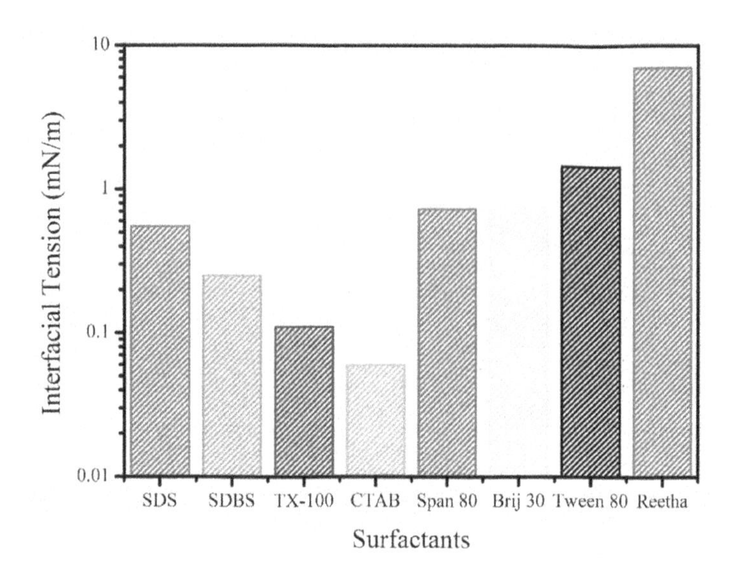

FIGURE 3.3
Effect of synthetic and natural surfactants towards IFT reduction for heavy crude oil (Assam, India).

results in the saturation of adsorption capacity, which produces a constant IFT profile [15]. It is only CTAB and TX-100 which could reduce the IFT to an ultra-low value of ~10^{-2}mN/m. Apart from ultra-low IFT, these two surfactants showed dynamic IFT variation which took almost 10 minutes for CTAB and 30 minutes for TX-100 to reach equilibrium. The lowest transient IFT values recorded by the surfactants CTAB and TX-100 are 6.2×10^{-3} and 3.1×10^{-2}mN/m, respectively. Similarly, equilibrium IFT values of 7.5×10^{-2} and 1.16×10^{-1}mN/m were detected for CTAB and TX-100 respectively. The remaining other surfactants do not show any transient behaviour in their IFT values [15].

The main reason for CTAB and TX-100 to result in minimum IFT with respect to other surfactants was because of the interaction between a hydrophilic and a hydrophobic group of surfactants, water and oil molecules at the oil–water interface [27]. The development in adsorption of surfactant molecules at the interface, packing of these surfactant molecules, large interfacial area, charge at the surface and surface viscosity additionally contributes towards ultra-low IFT [27–30]. The ultra-low IFT can further be achieved for surfactant retaining a special structure [27,31]. An interesting behaviour was observed while measuring the dynamic IFT for the surfactant CTAB only. The oil layer or drop so formed in the capillary tube undergoes elongation with time and breaks off without affecting the IFT as the L/D ratio was maintained in the range of ≥4. This phenomenon of time-dependent oil break might have resulted in better emulsification capacity for the surfactant CTAB.

3.3.2 Synergy of Emulsification and IFT

The reduction in IFT between the aqueous chemical phase and crude oil is expected to emulsify the crude oil. Therefore, the emulsification test was executed at a higher surfactant concentration of 0.3 wt%, which is much above the CMC values for the surfactants considered. The surfactant monomers above CMC values formed micelles which solubilize the oil or water forming either W/O emulsion or O/W emulsions.

Figure 3.4 represents a schematic diagram which clearly demonstrates the relation between IFT and emulsion formation. The emulsification capacity of crude oil increases and is inversely proportional to IFT or directly proportional to IFT reduction. In our published article [15], we observe similar behaviour while investigating the emulsion behaviour of all the above surfactants. It was found that the cationic surfactant CTAB which showed the lowest IFT was successful in emulsifying the crude oil to a greater extent in comparison to other surfactants. Moreover, the emulsion stability for CTAB was highest and the stability continued even after observing for 30 days. The stability of emulsion produced using surfactants depends on the electrostatic repulsion and steric hindrance at the oil–water interface, which decides the rate of coalescence of the oil droplets [32]. The emulsion extent or capacity followed the following order: CTAB > TX-100 > SDBS > SDS > Span 80 > Brij 30 > Titriplex III > Tween 80.

The IFT reduction and emulsification formation depend widely on the type of surfactant, its concentration, salinity, temperature and composition of crude oil. Therefore for the chosen system, we tried to develop a co-relation between IFT reduction and emulsification extent by plotting a linear graph for which R^2 was found to be 0.96. The relationship was obtained with the exclusion of two anionic surfactants Tween 80 and Titriplex III. Tween 80 during the process showed no sign of emulsification whereas with Titriplex III,

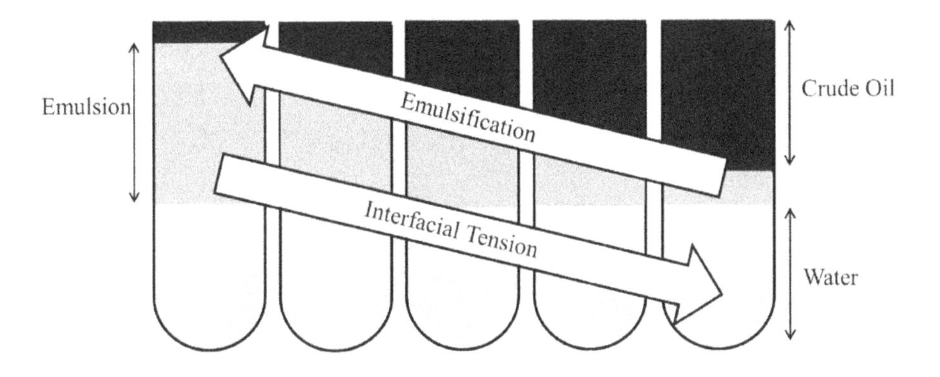

FIGURE 3.4
Schematic diagram representing the relationship between IFT values and emulsification capacity.

the IFT could not be measured [15]. Hence, the relationship obtained clearly showed or validated the IFT reduction phenomena as one of the selection or screening criteria for surfactant in the current system.

3.3.3 Thermal Stability of Surfactants

The temperature and composition of crude oil vary as we move from one oil field to another. Surfactants when injected in such reservoirs must be stable to promote successful chemical flooding. Hence, in order to evaluate the stability of potential surfactants (CTAB and TX-100), a thermal stability test has been executed. The methods used to validate the thermal stability of surfactants include short-time high-temperature exposure and long-time ageing of surfactant at 90°C for 10 days. The overall characterization techniques adopted includes thermogravimetric analysis (TGA) and Fourier Transform Infrared Spectroscopy (FTIR). Additionally, to further confirm the stability of surfactants, IFT measurements before and after thermal ageing of the surfactants were compared and reported.

Figure 3.5a–c represents the TGA, FTIR and IFT graph, respectively, of the virgin surfactants and aged surfactants. The TGA data showed the onset of

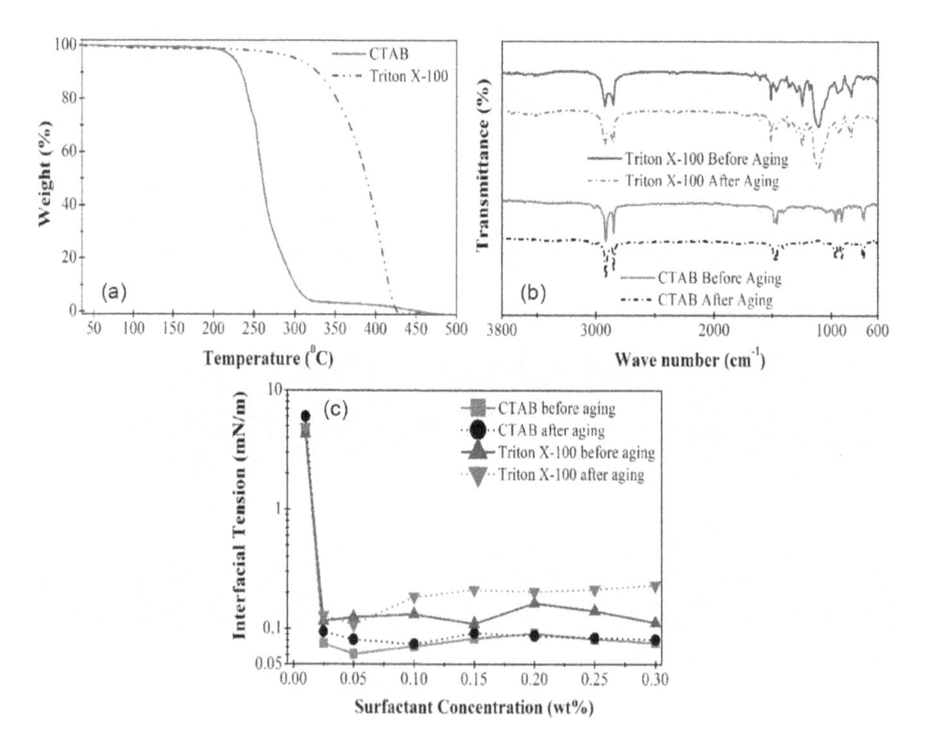

FIGURE 3.5
Detection of thermal stability of surfactants with various characterization techniques (a) thermogravimetric analysis, (b) FTIR spectrum and (c) interfacial tension studies.

surfactant degradation at 210°C for CTAB and 280°C for TX-100 (Figure 3.5a). TGA is thermal exposure for a short duration, whereas the surfactants when injected in reservoirs residue from several weeks to months. Therefore to incorporate the reservoirs scenario, thermal ageing of the surfactant was implemented. The FTIR spectrum of the raw surfactants and thermally aged samples confirmed no decomposition or reaction of surfactants as the functional groups detected with both the surfactant samples were more or less the same (Figure 3.5b). Moreover, the equilibrium IFT value obtained for both the selected surfactants (aged and non-aged) remained almost the same with a negligible variation. Therefore, the overall data obtained from different methods clearly demonstrated the stability of surfactants for most of the reservoirs bearing residual crude oil.

3.4 Formulation of Optimal Surfactant Composition

3.4.1 Dynamic IFT of Combined Surfactants

The efficient surfactants (CTAB and TX-100) were further mixed to detect the ultra-low IFT and understand the synergistic behaviour. We observed that the efficient surfactants were successful in reducing the IFT at a small concentration (0.025 wt%). Hence, the combined mixture of the surfactants (CTAB and TX-100) was formed by varying the concentration of the surfactants in the range of 0.01–0.05 wt% as shown in Table 3.1. The investigation reported a minimum transient IFT value of ~1×10^{-3} mN/m at 0.06 wt% (0.05 wt% CTAB+0.01 wt% TX-100) and 0.1 wt% (0.05 wt% CATB+0.05 wt% TX-100). However, at a concentration of 0.1 wt%, the lowest transient IFT was

TABLE 3.1

IFT Values between Heavy Crude Oil and Efficient Surfactant Mixture Solution at Different Surfactant Concentrations

Sr. No.	Surfactant (wt %) CTAB and TX-100		Time to Reach Lowest Transient IFT (minutes)	Lowest Transient IFT (mN/m)	Equilibrium IFT (mN/m)
1	0.01	0.01	15	1×10^{-2}	7.8×10^{-2}
2	0.01	0.025	10	5.4×10^{-3}	9.1×10^{-2}
3	0.01	0.05	10	8.1×10^{-3}	9.6×10^{-2}
4	0.025	0.01	5	9.1×10^{-3}	9.4×10^{-2}
5	0.025	0.025	5	1.1×10^{-2}	8.7×10^{-2}
6	0.025	0.05	5	1.4×10^{-2}	9.7×10^{-2}
7	0.05	0.01	5	1.8×10^{-3}	4.7×10^{-2}
8	0.05	0.025	3	7.4×10^{-3}	8.5×10^{-2}
9	0.05	0.05	3	1.4×10^{-3}	3.7×10^{-2}

detected in only 3 minutes and hence 0.1 wt% was selected as the optimum surfactant mixture value [15]. The reduction in dynamic IFT in less time will promote oil mobilization as the surfactant is expected to undergo an adsorption or dilution effect with time [32]. Moreover, the efficient surfactant mixture showed synergy bond interaction (cationic–non-ionic) at the interface, which results in the lowest IFT value of 10^{-2} mN/m [33].

3.4.2 Influence of Temperature and Salinity on the Optimum Surfactant Composition

As IFT reduction is a function of temperature and salinity, the efficient surfactant mixture was examined for IFT values at various temperature and salinity. The behaviour of IFT between heavy crude oil and optimum surfactant mixture with temperature showed a similar pattern as detected in alkali–oil chemistry. The IFT at room temperature was minimum (10^{-2} mN/m), which then starts to enhance with temperature (10^{-1} mN/m at 80°C). The molecule of surfactant diffuses at the interface, which results in a stable film that assists in IFT reduction. Moreover, as the temperature shoots up, the stable film undergoes destabilization, which thereby increases the IFT [23,24]. Moreover, the process of adsorption is exothermic in nature and hence hike in temperature of the system will hinder molecular diffusion of surfactant at the oil–water interface, which ultimately destabilizes the stable film.

The impact of salinity on the IFT behaviour between heavy crude oil and optimum surfactant mixture (S_M=0.05 wt % CTAB+0.05 wt % TX-100) and their corresponding emulsification formation capacity is depicted in Figure 3.6. The equilibrium IFT of the system reduces from 10^{-2} to ~10^{-3} mN/m as the concentration of the salinity approaches 10 wt% and on further increasing the salinity to 20 wt%, the equilibrium IFT reaches an ultra-low value of 3.2×10^{-3} mN/m. The main reason for IFT reduction at higher salinity was due to the increase in the quantity of adsorbed in-situ surfactant/soap at the oil–water interface [25,26,34]. In addition, the interaction of bond at the interface is improved by the dissolution activity of the in-situ surfactant/soap at the interface thereby minimizing the IFT [35].

While investigating the dynamic IFT behaviour we observed an important phenomenon of oil layer breaking. The oil drop is injected inside the capillary at certain rpm during which the L/D ratio is ≥4. After the oil is injected, the oil drop undergoes expansion or elongation inside the capillary tube with duration and finally breaks apart. The breaked oil drop also maintained the same L/D ratio and it was interesting to observe the corresponding emulsion behaviour under such a scenario.

An inverse proportional relationship was observed between emulsification capacity and oil layer break time or IFT reduction. Additionally, the oil break time, IFT reduction and emulsification are governed by the salinity of the system (Figure 3.6). We detected maximum emulsification when the system is free from any saline impact (formation water of 4445 ppm with zero salinity)

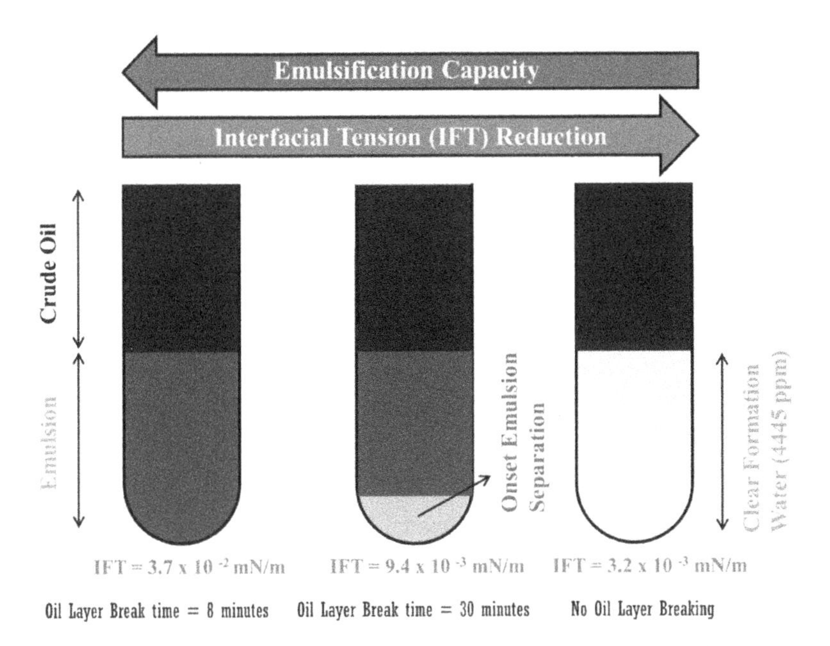

FIGURE 3.6
The impacts of salinity on IFT, oil layer break time and emulsification capacity for heavy crude oil (Assam, India).

and under such conditions, the oil layer break time was 8 minutes with an IFT of 10^{-2}mN/m. As salinity increases to 10wt%, the oil break time approaches 30 minutes and the IFT reduction enhances to ~10^{-3}mN/m followed by the onset of emulsion separation. Moreover, at extreme salinity (15–20wt%), no oil drop break could be seen and the emulsion formed was drastically reduced. The deviation in emulsification capacity and oil break time with salinity could be due to the variation in oil–water solubility ratio [36]. When the salinity of the system is low, more quantity of surfactant molecules lies at the oil–water interface which assists for early oil break time and enhanced emulsification capacity. On increasing the salinity, more Na^+ ions are associated in the system, which tends to remove the surfactant molecules from the interface, thus severely affecting the oil break time and emulsification extent. Additionally, the hydrophilic and lipophilic balance of the surfactant and the crude properties are also responsible for the mechanism of emulsification [37]. An in-depth investigation on the emulsification extent (emulsion type) with salinity (optimum salinity) can be found by observing the phase behaviour studies [33,36].

3.4.3 Adsorption Behaviour of Optimum Surfactant Composition

The chemical while flowing in the porous media undergoes adsorptions due to the ionic attraction of the rock minerals. Therefore, the understanding of the adsorption behaviour is important in order to incorporate the dilution effect of slug while forming the optimum slug. Here, in this work, we evaluated the adsorption quantity of TX-100 on carbonate Berea core at its optimum concentration of 0.05 wt%. The adsorbed amount was 0.35 mg/g and such a low adsorbed quantity could be due to the ionic nature of the non-ionic surfactant and rock core. Furthermore, the enabling of a steric hindrance at the surface interface during the surfactant adsorption process additionally accounts for lower adsorption [38,39].

The adsorption of CTAB (cationic surfactant) on the rock surface cannot be equal to or more than TX-100 because CTAB and rock surface possess similar charge. The same charge between them will result in repulsive force, thereby ultimately undergoing minimum adsorption. However, as per our understanding of the ionic charge attraction, it can be concluded that the maximum adsorption that the system will follow cannot exceed more than 0.7 mg/g. Therefore, if core flooding experiments are performed for this system, the adsorption is expected to vary between 0.35 and 0.7 mg/g. If alkali is introduced in the system, then the adsorption quantity will further reduce as alkali will act as a sacrifice agent and block the active adsorption sites of the rock surface. A detailed investigation on surfactant adsorption with rock surfaces is explained in Chapter 4.

3.5 IFT between Alkali–Surfactant Combinations

The synergy of the alkali–surfactant mixture was investigated by mixing alkali (0.6% NaOH) and surfactants (0.05% CTAB+0.05 TX-100) at their optimum values. The alkali–surfactant mixture solutions after reacting with heavy crude oil resulted in an ultra-low IFT value of around 6.7×10^{-3} mN/m. The time required to reach this equilibrium ultra-low IFT was around 40 minutes. Though we were successful in reaching the ultra-low IFT region, the emulsion produced in such a scenario was severely affected. Hence, emulsification behaviour at the optimum concentration for all the chemicals (alkali, surfactant, surfactant mixture and alkali–surfactant mixture) was investigated.

Figure 3.7 depicts a schematic diagram representing the emulsion behaviour and IFT values for all chemicals at their optimum concentrations values. The alkali system was successful in reducing the IFT; however, the emulsion obtained was almost negligible. As emulsification and IFT reduction mechanisms together assist towards higher oil recovery, the current alkali system (0.6% NaOH) is expected to provide lower oil recovery. The emulsification was much better when the efficient surfactant mixture (0.05% CTAB+0.05% TX-100) was mixed with the heavy crude; however, the stability of the emulsion was drastically reduced within 24 hours. Similar emulsification behaviour was observed with the alkali–surfactant mixture system but the stability

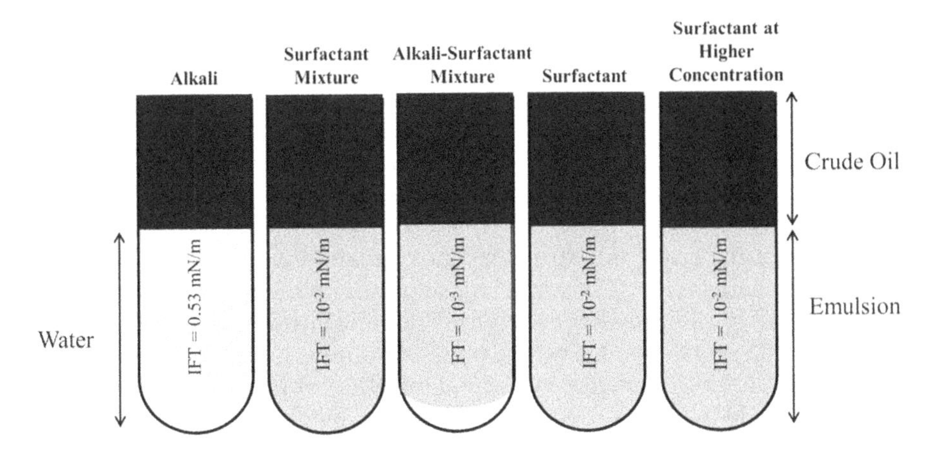

FIGURE 3.7
Schematic diagram representing reduced IFT value and its corresponding emulsification for alkali (0.6% NaOH), surfactant mixture (0.05% CTAB+0.05% TX-100), alkali–surfactant mixture (0.6% NaOH+0.05% CTAB+0.05% TX-100) and individual surfactant CTAB.

of the emulsion could not extend beyond 180 minutes. The main reason for lower emulsion stability could be due to the replacement of surfactant molecules by Na^+ ions (in-situ soap) in the oil–water system. Therefore to enhance the emulsion formation and its stability, individual surfactant CTAB was introduced. A strong emulsion was observed when the concentration of CTAB reaches 0.2 wt% and such emulsions were stable even after 24 hours. Moreover, the formation of a strong emulsion can severely affect the recovery rate due to poor displacement efficiency and such confirmation can be observed by performing core flooding experiments.

3.6 Wettability Alteration with Different Chemicals

The change in the initial wettability of the reservoir rock towards more favourable wettability can enhance the oil recovery process by inducing movement of the residual oil inside the pores [40–43]. The wettability change can be estimated by measuring the contact angle between a liquid drop and a rock surface. Therefore, the contact angle between carbonate Berea cores and chemicals like alkali, surfactant mixture and alkali–surfactant mixtures at their optimum values were estimated to identify the participation of the wettability alteration mechanism. The initial wettability of the Berea rock was evaluated by measuring the contact angle between the solid carbonate rock and formation water (4445 ppm). The contact angle at the beginning was around 80° which then reaches 51° as the final value. The obtained value clearly shows the initial wettability of the reservoir to be intermediate wet [15,40]. The initial contact angles for all the chemical solutions at their optimum values were in the range of 47°–28°. Moreover, this contact angle reduces to around 10° with time, which represents the induction of wettability change in the system from intermediate wet to favourable water wet. The active role of alkali and formation water towards ionization of the acid group of the crude oil introduces the change in wettability by the ion-binding mechanism [44]. The surfactant solutions on the other hand produce ion pairs due to the interaction between cationic surfactant molecules and the carboxylic group of oil. The ion pairs then undergo diffusion in the oil phase as well as in the aqueous surfactant micelles, which causes the water to penetrate inside the pores of the rock thus changing the wettability to desirable water wet [45].

3.7 Berea Core Flooding for Oil Recovery

The cumulative oil recoveries obtained using chemicals (alkali, surfactant, surfactant mixture and alkali-surfactant mixture) at their optimum concentration are illustrated in Figure 3.8 and Table 3.2. The initial water flooding was executed till 2 PV and then 0.5 PV chemical slug flowed by 1.5 PV chase water flooding was performed to estimate the cumulative oil recovery

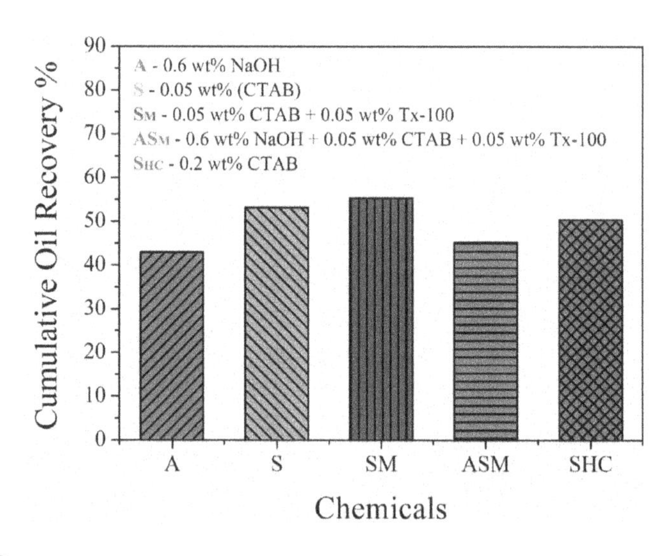

FIGURE 3.8
Cumulative heavy oil recovery obtained using alkali, surfactant, surfactant mixture and alkali–surfactant mixtures for carbonate reservoir.

TABLE 3.2

Summary of Core Flooding Predicting Oil Recovery for Carbonate Reservoir Using Various Chemical Combinations

Porosity (%)	Brine Permeability (mD)	Initial Oil Saturation (%)	Chemical Formulation (wt%)	Water Flooding Recovery (%)	Chemical Recovery (IOIP %)	Total Oil Recovery (%)	Residual Oil Recovery (ROIP %)
20.8	90.3	80.1	0.6% NaOH (A)	30.16	12.79	42.95	18.31
20.3	92.9	79.9	0.025% CTAB (S)	29.52	16.87	46.39	23.94
19.6	71.9	82.5	0.05% CTAB	30.57	22.58	53.15	32.52
20.9	74.1	79.9	0.1% CTAB	30.13	21.91	52.04	31.36
19.4	66.7	79.4	0.2% CTAB	30.11	20.24	50.35	28.96
20.3	49.2	79.1	0.05% CTAB+0.05% TX-100 (S_M)	30.72	24.58	55.30	35.48
20.2	82.6	79.6	0.1% CTAB+0.1% TX-100	30.26	21.75	52.01	31.87
20.3	84.3	80.3	0.6% NaOH+0.05% CTAB+0.05% TX-100	30.78	14.46	45.24	20.89
19.9[a]	68.4	80.1	0.6% NaOH+0.05% CTAB+0.05% TX-100 (AS_M); 0.05% CTAB+0.05% TX-100 (S_M)	31.07	18.65	49.72	27.06

[a] Dual slug injected – First chemical slug of 0.5 PV followed by second chemical slug of 0.5 PV and then 1 PV chase water flooding.

[15,46–48]. The total oil recovered using alkali flooding was around 43% and the reason for such low recovery was the negligible emulsification effect of the crude oil. The cumulative oil recovery enhances to around 46.5% when alkali is replaced by 0.025 wt% CTAB and finally reaches 53% as the quantity of surfactant approaches 0.05 wt%. The overall oil recovery by chemical flooding was maximized to almost 55.5% when 0.05 wt% CTAB+0.05 wt% TX-100 surfactant mixture was used. The enhanced oil recovery with surfactant and surfactant mixture was due to the formation of ultra-low IFT and effective emulsion, which promotes better sweep efficiency by reducing viscous fingering. In addition, the maximized oil recovery can be further connected to the time-dependent achievement of lowest or minimum transient/dynamic IFT [15,32].

The alkali surfactant flooding (0.06% NaOH+0.05 wt% CTAB+0.05 wt% TX-100) on the other hand showed a drastic reduction in oil recovery. The oil recovery was around 45% and such low recovery was due to the effect of emulsion formation capacity and its stability. Additionally, to validate the role of the emulsification mechanism for the alkali–surfactant mixture system, an additional slug of 0.5 PV surfactant mixtures (0.05 wt% CTAB+0.05 wt% TX-100) was injected right after the alkali–surfactant mixture injection followed by 1 PV chase water flooding [15]. Thus, in doing so, the oil recovery was increased by another 4%–5%, which clearly indicated the dominant role of emulsification mechanisms towards the oil recovery factor (Table 3.2). Moreover, emulsification further decides or controls the sweep efficiency by diverting the aqueous phase to the uncovered area of the reservoir ultimately enhanced the displacement efficiency. However, a severe decrease in oil recovery can be observed due to the detrimental effect of emulsion on sweep efficiency. We observed such impact when the individual concentration of surfactant (CTAB) was increased from 0.05 to 0.2 wt%. The oil recovery reduces from 53 to almost 50% under such a scenario as the emulsion formed was too strong and is difficult to displace or flow the oil through the porous media thus reducing the sweep efficiency [32,49]. Hence, for this type of system, we can add polymer to displace the strong emulsion as well as oil for improving the recovery factor (polymer flooding is explained in Chapter 5).

3.8 Conclusions

The study for formulation of optimum chemical concentration for various alkalis, surfactants, surfactant mixtures and alkali–surfactant mixtures was performed. The screening of chemicals was initially performed based on the reduced equilibrium IFT values. Apart from equilibrium IFT, transient/dynamic IFT values were also scrutinized to identify the optimum values. The effects of temperature and salinity on the IFT values were explored to describe or understand the complexities of the system. Moreover, the emulsification mechanism was recognized to understand the improved oil recovery factor. The emulsification factor plays a dominant role in deciding the displacement efficiency, which directly controls the overall oil recovery from the reservoirs. Moreover, the drastic reduction in the contact angle demonstrated the alteration in the wettability of the carbonate reservoir from intermediate wet to desirable water wet. Alkali flooding at the optimum value of 0.6 wt% showed an additional oil recovery of 13% and such lower recovery could be due to the absence of an emulsification mechanism. The additional oil recovery enhances to 22.5% and 25% at 0.05 wt% CTAB (surfactant) and 0.05 wt% CTAB + 0.05 wt% TX-100 (surfactant mixture) respectively due to the emulsion effect. Thus, the results clearly portray the dominant role of emulsification towards higher oil recovery. Additionally, the alkali–surfactant mixture was not much effective for higher oil recovery, which is due to reduced emulsification and its lower stability. Moreover, the oil recovery also depends on the quality of emulsion as the formation of strong emulsion leads to poor mobility or poor oil displacement thus reducing the oil recovery factor.

References

1. J. J. Sheng, Status of alkaline-surfactant flooding, *Polymer Sciences*, vol. 1, pp. 1–7, 2015. doi: 10.21767/2471-9935.100006.
2. J. J. Sheng, Alkaline-surfactant flooding, in *Enhanced Oil Recovery Field Case Studies*. Elsevier, Amsterdam, pp. 179–188, 2013.
3. M. Dong, S. Ma, and Q. Liu, Enhanced heavy oil recovery through interfacial instability: A study of chemical flooding for Brintnell heavy oil, *Fuel*, vol. 88, pp. 1049–1056, 2009.
4. Q. Liu, M. Dong, X. Yue, and J. Hou, Synergy of alkali and surfactant in emulsification of heavy oil in brine, *Colloids and Surfaces A: Physicochemical and Engineering Aspects*, vol. 273, pp. 219–228, 2006.
5. L. Chen, G. Zhang, J. Ge, P. Jiang, J. Tang, and Y. Liu, Research of the heavy oil displacement mechanism by using alkaline/surfactant flooding system, *Colloids and Surfaces A: Physicochemical and Engineering Aspects*, vol. 434, pp. 63–71, 2013.
6. H. Pei, G. Zhang, J. Ge, M. Tang, and Y. Zheng, Comparative effectiveness of alkaline flooding and alkaline–surfactant flooding for improved heavy-oil recovery, *Energy and Fuels*, vol. 26, pp. 2911–2919, 2012.
7. J. L. Bryan and A. Kantzas, Enhanced heavy-oil recovery by alkali-surfactant flooding, SPE-110738-MS, *in SPE Annual Technical Conference and Exhibition*, 11–14 November, Anaheim, California, 2007.
8. H. H. Pei, G. C. Zhang, J. J. Ge, L. Ding, M. G. Tang, and Y. F. Zheng, A comparative study of alkaline flooding and alkaline/surfactant flooding for zhuangxi heavy oil, SPE 146852, *in SPE Heavy Oil Conference*, 12–14 June, Calgary, Alberta, Canada, 2012.

9. R. Saha, A. Sharma, R. Uppaluri, and P. Tiwari, Interfacial interaction and emulsification of crude oil to enhance oil recovery, *International Journal of Oil, Gas and Coal Technology*, vol. 22, pp. 1–15, 2019.
10. P. O. Roehl and P. W. Choquette, *Carbonate Petroleum Reservoirs*. Springer-Verlag, New York, 1985.
11. G. V. Chilingar and T. F. Yen, Some notes on wettability and relative permeabilities of carbonate reservoir rocks, II, *Energy Sources*, vol. 7, pp. 67–75, 1983.
12. M. Bai, Z. Zhang, X. Cui, and K. Song, Studies of injection parameters for chemical flooding in carbonate reservoirs, *Renewable and Sustainable Energy Reviews*, vol. 75, pp. 1464–1471, 2017.
13. E. Hosseini, F. Hajivand, and R. Tahmasebi, The effect of alkaline–surfactant on the wettability, relative permeability and oil recovery of carbonate reservoir rock: Experimental investigation, *Journal of Petroleum Exploration and Production Technology*, vol. 9, pp. 2877–2891, 2019.
14. K. Mohan, Alkaline surfactant flooding for tight carbonate reservoirs, SPE-129516-STU, *in SPE Annual Technical Conference and Exhibition*, New Orleans, Louisiana, 4–7 October, 2009.
15. R. Saha, R. Uppaluri, and P. Tiwari, Effects of interfacial tension, oil layer break time, emulsification and wettability alteration on oil recovery for carbonate reservoirs, *Colloids and Surfaces A*, vol. 559, pp. 92–103, 2018.
16. H. A. Nasr-El-Din and K. C. Taylor, Dynamic interfacial tension of crude oil/alkali/surfactant systems, *Colloids and Surfaces*, vol. 66, pp. 23–37, 1992.
17. L. Mei-qin, X. Xue-qin, L. Jing, Z. Hua, L. Ming-yuan, and D. Zhao-xia, Interfacial properties of Daqing crude oil-alkaline system, *Petroleum Science*, vol. 8, pp. 93–98, 2011.
18. J. Ge, A. Feng, G. Zhang, P. Jiang, H. Pei, R. Li, et al., Study of the factors influencing alkaline flooding in heavy-oil reservoirs, *Energy and Fuels*, vol. 26, pp. 2875–2882, 2012.
19. H. Pei, G. Zhang, J. Ge, L. Jin, and C. Ma, Potential of alkaline flooding to enhance heavy oil recovery through water-in-oil emulsification, *Fuel*, vol. 104, pp. 284–293, 2013.
20. H. Y. Jennings, C. E. Johnson, and C. D. McAuliffe, A caustic waterflooding process for heavy oils, *Journal of Petroleum Technology*, vol. 26, pp. 1344–1352, 1974.
21. J. Rudin and D. T. Wasan, Mechanisms for lowering of interfacial tension in alkali acidic oil system 1. Experimental studies, *Colloids and Surfaces*, vol. 68, pp. 67–79, 1992.
22. J. Rudin and D. T. Wasan, Mechanisms for lowering of interfacial tension in alkali/acidic oil systems 2. Theoretical studies, *Colloids and Surfaces*, vol. 68, pp. 81–94, 1992.
23. D. L. Flock, T. H. Le, and J. P. Gibeau, The effect of temperature on the interfacial tension of heavy crude oils using the pendent drop apparatus, *Journal of Canadian Petroleum Technology*, vol. 25, pp. 72–77, 1986.
24. O. S. Hjelmeland and L. E. Larrondo, Experimental investigation of the effects of temperature, pressure, and crude oil composition on interfacial properties, *SPE Reservoir Engineering*, vol. 1, pp. 321–328, 1986.
25. A. J. Prosser and E. I. Franses, Adsorption and surface tension of ionic surfactants at the air–water interface: Review and evaluation of equilibrium models, *Colloids and Surfaces A: Physicochemical and Engineering Aspects*, vol. 178, pp. 1–40, 2001.

26. J. Jiravivitpanya, K. Maneeintr, and T. Boonpramote, Experiment on measurement of interfacial tension for subsurface conditions of light oil from Thailand, *in MATEC Web of Conferences, ICCME,* Phuket, Thailand, 2016.
27. H.-R. Li, Z.-Q. Li, X.-W. Song, C.-B. Li, L.-L. Guo, L. Zhang, et al., Effect of organic alkalis on interfacial tensions of surfactant/polymer solutions against hydrocarbons, *Energy Fuels,* vol. 29, pp. 459–466, 2015.
28. L. Fu, G. Zhang, J. Ge, K. Liao, H. Pei, P. Jiang, et al., Study on organic alkali-surfactant-polymer flooding for enhanced ordinary heavy oil recovery, *Colloids and Surfaces A: Physicochemical and Engineering Aspects,* vol. 508, pp. 230–239, 2016.
29. J. R. Kanicky, J.-C. Lopez-Montilla, S. Pandey, and D. O. Shah, Surface chemistry in the petroleum industry, in Holmberg, K., (Ed.), *Handbook of Applied Surface and Colloid Chemistry.* John Wiley & Sons, Ltd, Gainesville, FL, pp. 251–267, 2001.
30. J. Rudin and D. T. Wasan, Mechanisms for lowering of interfacial tension in alkali/acidic oil systems: Effect of added surfactant, *Industrial & Engineering Chemistry Research,* vol. 31, pp. 1899–1906, 1992.
31. L. Zhang, L. Luo, S. Zhao, and J. Y. Yu, *Ultra Low Interfacial Tension and Interfacial Dilational Properties Related to Enhanced Oil Recovery.* Nova Science Publishers, New York, 2008.
32. C. D. Yuan, W. F. Pu, X. C. Wang, L. Sun, Y. C. Zhang, and S. Cheng, Effects of interfacial tension, emulsification, and surfactant concentration on oil recovery in surfactant flooding process for high temperature and high salinity reservoirs, *Energy and Fuels,* vol. 29, pp. 6165–6176, 2015.
33. D. W. Green and G. P. Willhite, *Enhanced Oil Recovery,* vol. 6. SPE Textbook Series. SPE, Richardson, TX, 1998.
34. A. Bera, A. Mandal, and B. B. Guha, Synergistic effect of surfactant and salt mixture on interfacial tension reduction between crude oil and water in enhanced oil recovery, *Journal of Chemical & Engineering Data,* vol. 59, pp. 89–96, 2014.
35. Z. Zhao, C. Bi, W. Qiao, Z. Li, and L. Cheng, Dynamic interfacial tension behavior of the novel surfactant solutions and Daqing crude oil, *Colloids and Surfaces A: Physicochemical and Engineering Aspects,* vol. 294, pp. 191–202, 2007.
36. A. Bera, K. Ojha, A. Mandal, and T. Kumar, Interfacial tension and phase behavior of surfactant-brine–oil system, *Colloids and Surfaces A: Physicochemical and Engineering Aspects,* vol. 383, pp. 114–119, 2011.
37. P. S. Piispanen, M. Persson, P. Claesson, and T. Norin, Surface properties of surfactants derived from natural products part 1: Syntheses and structure/property relationships-solubility and emulsification, *Journal of Surfactants and Detergents,* vol. 7, pp. 147–159, 2004.
38. M. A. Ahmadi, S. Zendehboudi, A. Shafiei, and L. James, Nonionic surfactant for enhanced oil recovery from carbonates: Adsorption kinetics and equilibrium, *Industrial and Engineering Chemistry Research,* vol. 51, pp. 9894–9905, 2012.
39. M. A. Ahmadi and S. R. Shadizadeh, Experimental investigation of a natural surfactant adsorption on shale-sandstone reservoir rocks: Static and dynamic conditions, *Fuel,* vol. 159, pp. 15–26, 2015.
40. S. Kumar and A. Mandal, Studies on interfacial behavior and wettability change phenomena by ionic and nonionic surfactants in presence of alkalis and salt for enhanced oil recovery, *Applied Surface Science,* vol. 372, pp. 42–51, 2016.
41. P. Kathel and K. K. Mohanty, Wettability alteration in a tight oil reservoir, *Energy and Fuels,* vol. 27, pp. 6460–6468, 2013.

42. C. S. Vijapurapu and D. N. Rao, Compositional effects of fluids on spreading, adhesion and wettability in porous media, *Colloids and Surfaces A: Physicochemical and Engineering Aspects,* vol. 241, pp. 335–342, 2004.
43. O. R. Wagner and R. O. Leach, Improving oil displacement efficiency by wettability adjustment, *Society of Petroleum Engineers,* vol. 216, pp. 65–72, 1959.
44. Q. Liu, M. Z. Dong, K. Asghari, and Y. Tu, Wettability alteration by magnesium ion binding in heavy oil/brine/chemical/sand systems: Analysis of hydration forces, *Natural Science,* vol. 2, pp. 450–456, 2010.
45. D. C. Standnes and T. Austad, Wettability alteration in chalk 2: Mechanism for wettability alteration from oil-wet to water-wet using surfactants, *Journal of Petroleum Science & Engineering,* vol. 28, pp. 123–143, 2000.
46. R. Saha, R. Uppaluri, and P. Tiwari, Impact of natural surfactant (Reetha),pPolymer (Xanthan Gum), and silica nanoparticles to enhance heavy crude oil recovery, *Energy and Fuels,* vol. 33, pp. 4225–4236, 2019.
47. R. Saha, R. V. S. Uppaluri, and P. Tiwari, Influence of emulsification, interfacial tension, wettability alteration and saponification on residual oil recovery by alkali flooding, *Journal of Industrial and Engineering Chemistry,* vol. 59, pp. 286–296, 2018.
48. R. Saha, R. Uppaluri, and P. Tiwari, Silica nanoparticle assisted polymer flooding of heavy crude oil: Emulsification, rheology, and wettability alteration characteristics, *Industrial and Engineering Chemistry Research,* vol. 57, pp. 6364–6376, 2018.
49. S. Cobos, M. S. Carvalho, and V. Alvarado, Flow of oil–water emulsions through a constricted capillary, *International Journal of Multiphase Flow,* vol. 35, pp. 507–515, 2009.

4

Surfactant Adsorption Characteristics on Reservoir Rock

4.1 Introduction

Surfactant flooding is a branch of chemical EOR having a great potential to enhance the recovery of residual oil [1–3]. The efficiency of surfactant flooding to improve the residual oil recovery has been validated by several authors [4–7]. Surfactant flooding is effective due to its ability to reduce the interfacial tension (ultra-low IFT) between oil and water and/or alter the wettability of the reservoir rock thus mobilizing the trapped residual oil [2,8–12]. A detailed investigation reported that oil recovery by surfactant flooding depends on different mechanisms like IFT reduction, wettability alteration, emulsification and improved sweep efficiency [7,13–16].

A surfactant molecule is amphiphilic in nature which means that it contains two parts, namely a hydrophilic head or polar group which is water loving and a hydrophobic tail or non-polar group which is hydrocarbon loving. Therefore, based on the polar head, surfactants that are employed for oil recovery applications are generally categorized as cationic, anionic, non-ionic and zwitterionic groups [10,17,18]. The other types include Gemini surfactants, natural surfactants, polymeric surfactants and viscoelastic surfactants [10]. The charges on the hydrophilic head group are negative for anionic surfactants and are classified as carboxylate ($RCOO^-$), sulfonate (RSO_3^-) or sulfate (RSO_4^-). If the charge of the head group is positive, then they are known as cationic surfactants ($R_4N^+Cl^-$). For non-ionic surfactants, the charge on the head group is neutral, whereas for zwitterionic groups, it bears both positive and negative charges [10,19–21]. Gemini surfactants are those surfactants that possess more than one hydrophobic tail and hydrophobic head that are chemically bonded by a spacer [10,22]. Natural surfactants are derived surfactants from natural sources like seeds or fruits (Reetha) [23,24] and trees (Z. spina-christie) [10,25–27]. A viscoelastic surfactant on the other hand retains desired properties like viscosity for favourable mobility and minimum IFT [10]. The macromolecule structures in which both hydrophilic and hydrophobic parts are detected are commonly referred to as polymeric surfactants [10,28].

Surfactant flooding is employed when polymer flooding may not be sufficient to reduce the residual oil saturation [12]. In fact, in several cases, it is used in a combined form like alkali–surfactant [3,29], surfactant–polymer [30] and alkali–surfactant–polymer flooding [19,31,32]. Recently, natural surfactants have been explored to compare their potential for residual oil recovery. They are effective in terms of IFT reduction, wettability alteration, achieving favourable mobility ratio and hence can be competitive with respect to synthetic surfactants [33,34]. Natural surfactants are cheap, easily available, eco-friendly and can reduce the residual oil saturation to an acceptable extent [23,25].

Surfactants when injected inside the oil reservoir interact with the rock surface and undergo adsorption characteristics, which are inevitable [1,25,35–37]. Hence, surfactant adsorption is one of the important criteria, which have to be evaluated in order to avoid dilution of the optimum chemical slug and at the same time, the economy of the process can be evaluated considering the cost of the surfactant. The factors which influence the adsorption quantity are the temperature, salinity, pH, adsorbent surface area [38–40] and the charge species between adsorbent and adsorbate [41,42]. Numerous studies on the adsorption of surfactant to the rock surface have been performed. This includes adsorption of different surfactants like synthetic [35,36,39,40,43,44] and natural surfactants [25,27,45–48] on the rock surface like reservoir rock, sandstone, Berea cores [27,35,39,42,47,49–53] and carbonate rock [37,54]. The studies fitted experimental adsorption data in different isotherm and kinetics models to understand the adsorption characteristic in detail.

An important investigation confirmed the relation between clay content of the adsorbent and the amount of surfactant adsorption using the CMC (critical micelle concentration) method [42]. In the study, different compositions of quartz sand and clay minerals were utilized to estimate the adsorption quantity of anionic and non-ionic surfactants. The adsorption was highest for montmorillonite and the order followed is montmorillonite $\gg$ illite $>$ kaolinite. However, more experimental findings and data are required to confirm the existing model.

Therefore in this chapter, we will focus on the adsorption behaviour of the selected surfactant for sandstone type Indian reservoir rock. As the reservoir rock is of sandstone nature, only anionic (sodium lignosulphonate and SDS – sodium dodecyl sulphate) and non-ionic (Triton X-100, Tween 80, Titriplex III, Brij 30 and Span 80) surfactants were screened based on the charge species. The surfactant was selected based on IFT reduction and different parameters that influence the adsorption quantity like rock mineralogy, salinity, temperature and so on were analysed. Finally, numerous isotherm and kinetic models were fitted to analyze and predict the adsorption characteristics.

4.2 Characterization of Rock Samples

Sandstone-type reservoir rock samples were collected from (Hapjan and Jorajan) one of the North-East states, Assam, India. Berea cores such as Carbon Tan, Gray Berea and Idaho Gray were procured from Kocurek Industries, Hard rock division, USA. The rock samples were characterized by XRD (X-ray powder diffraction), EDX (energy-dispersive X-ray spectroscopy) and BET (Brunauer–Emmett–Teller) analytical techniques to estimate the rock mineralogy, charge nature and surface or pore properties respectively. The compositional analysis of the rock samples (Assam field and Berea cores) was illustrated by XRD and is majorly composed of quartz (36%–80%), illite (5%–33%), feldspar (3%–15%), kaolinite (2%–11%), dolomite (1%–40%) and montmorillonite (0.1%–4%) [55]. The EDX analysis of Hapjan rock (Assam oil field) demonstrates that the majority of the components present in the sample are oxygen (53.1%) and silica (31.2%) with minor quantities of aluminium (6.4%), iron (4.4%), magnesium (3.2%), calcium (0.6%), potassium (0.6%) and sodium (0.5%). Similarly, BET data depict that the average pore radius of the Hapjan rock is 6.99 nm with a surface area of 4.97 m^2/g, a total pore volume of 0.017 cc/g, pores of slit shape and the existence of several micropores, macropores and mesopores [55].

4.3 Interfacial Tension between Crude Oil and Surfactant

4.3.1 IFT Behaviour Using Different Surfactants

The reduction in interfacial tension (IFT) between surfactants (two anionic and five non-ionic groups) and Assam crude oil of light to moderate category was studied for screening the surfactants as depicted in Figure 4.1. The initial choices of surfactants were considered based on the type of reservoir rock, which is sandstone in nature and hence cationic surfactants were not employed due to opposite charge species. The IFT reduction was significant for Triton X-100 and was found to be 0.22 mN/m. The IFT reduction for other surfactants followed the following order: sodium dodecyl sulphate (0.78 mN/m) < span 80 (0.98 mN/m) < sodium lignosulphonate (2.14 mN/m) < Tween 80 (2.82 mN/m) < Brij 30 (3.12 mN/m) [55]. The remaining surfactant named Titriplex III does not form any oil droplets during its analysis and hence its IFT could not be measured. Therefore, based on the IFT reduction behaviour for all the surfactants, only Triton X-100 was selected for further experiments.

The reduction in IFT was mainly observed due to the adsorption of surfactant molecules at the oil–water interface, which reduces the dissimilarities

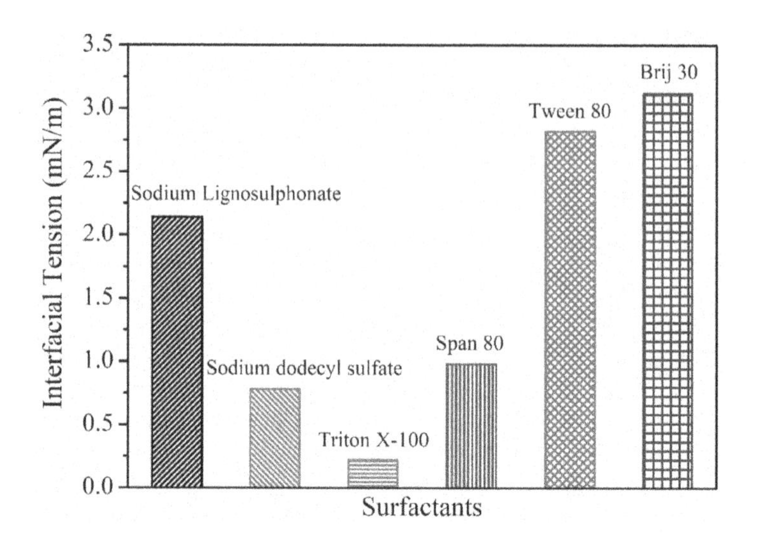

FIGURE 4.1
Interfacial tension (IFT) behaviour between light to moderate Assam (Indian) crude oil and different surfactants.

between the water side and oil side molecules at the interface [56]. The interaction between the hydrophilic and hydrophobic group of the water, surfactant and oil molecules can induce ultra-low IFT and can be introduced with the surfactant bearing special structures [56,57]. The other probable cause for IFT reduction as reported by several authors could be greater adsorption of surfactant molecules at the interface and their adsorption strength, packing of surfactant molecules at the interface, large interaction energy at the interface, high interfacial activity unionized acid species, surface charge, surface viscosity and surface elasticity [31,56,58,59].

4.3.2 IFT Behaviour with Formation Water

The equilibrium IFT between crude oil and the selected surfactant (Triton X-100) was measured using both formation water (~4500 ppm salinity) and millipore water as an aqueous phase [55]. Figure 4.2 shows the impact of salinity on the IFT behaviour as the concentration of surfactant varied up to 1 wt%. The IFT value of Assam crude oil (India) with millipore water decreased from 24.3 mN/m to a lowest value of 2.29 mN/m as the concentration of surfactant increased from 0 wt% to 0.02 wt% respectively and beyond which the reduction of IFT was negligible. A similar pattern was observed with formation water where the IFT reduced from 16.3 mN/m at 0 wt% surfactant to 0.22 mN/m at a concentration of 0.02 wt% and a negligible reduction thereafter. This shows that IFT reduction is significant in the presence of saline or sea water and similar results were also observed by other studies as depicted in Figure 4.2. The reason for IFT reduction was the accumulation of

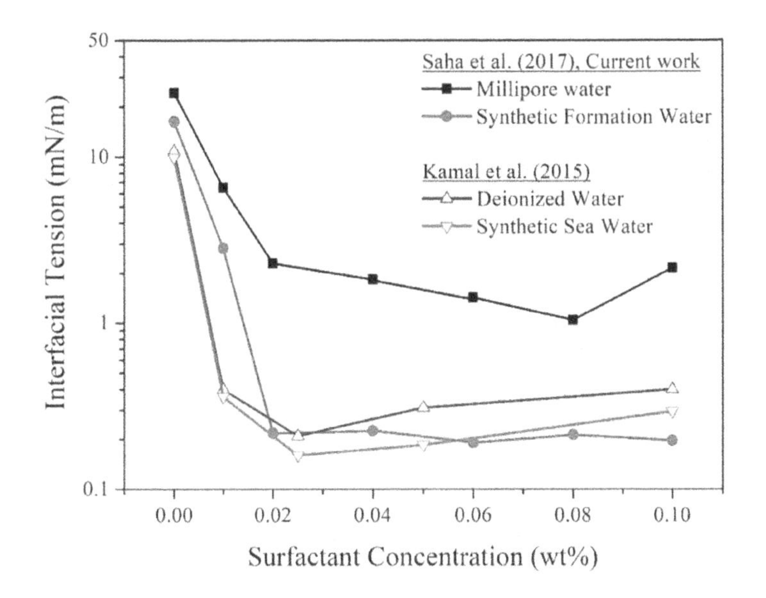

FIGURE 4.2
IFT behaviour with formation/sea water and millipore/deionized water.

surfactant molecules at the interface [60,61]. In the presence of salt, the extent of in-situ soap grows at the interface and hence reduces the IFT further to the lowest value. A synergy between salt in the formation water and surfactant molecules was observed [62]. The synergy interaction influences the double electrode layer and results in a thin layer formed at the oil–water interface by the arrangement of in-situ soap and surfactant molecules, producing the lowest IFT [63–65].

4.4 Thermal Stability of Surfactants

Surfactant stability is one of the most important parameters that must be evaluated to determine its potential for extreme reservoir conditions. Therefore, to validate surfactant stability, few characterizations like TGA (thermogravimetric analysis), NMR (nuclear magnetic resonance) and FTIR (Fourier transform infrared spectroscopy) followed by IFT analysis of the aged and non-aged Triton X-100 surfactant samples were carried out. Figure 4.3 shows the TGA, FTIR and NMR spectra of Triton X-100. The TGA spectrum of raw Triton X-100 revealed the initiation of surfactant degradation at 305°C and complete degradation at 425°C (Figure 4.3a). Hence, the thermal stability of the chosen surfactant was validated as the temperature for majority of the oil reservoirs falls below the degradation limit as detected above. As TGA is

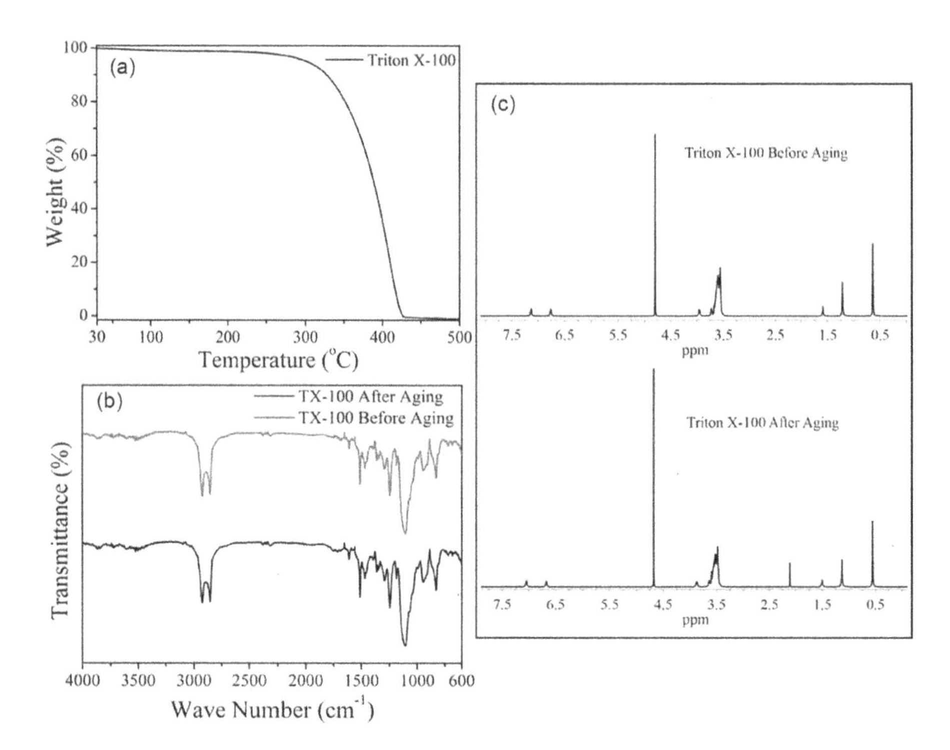

FIGURE 4.3
Thermal stability analysis of surfactant Triton X-100 by (a) TGA, (b) FTIR and (c) ^{1}H NMR.

a short period temperature exposure and duration of surfactants residence inside the reservoirs is quite high, ageing of the surfactant was executed. The surfactant was thermally aged in an oven for a period of 10 days at 90°C. The aged sample was then further analysed with NMR, FTIR and IFT studies to detect the degradation of surfactant if any.

The FTIR and NMR graphs of the aged and non-aged surfactant samples are illustrated in Figure 4.3b and c. The spectra of FTIR and NMR showed no variation in their functional group and peak or peak position, respectively; this represents the stability of the surfactant with respect to chemical analysis [66].

4.4.1 IFT of Aged and Non-Aged Surfactant Samples

The IFT values between crude oil and surfactant samples (aged and non-aged Triton X-100) were investigated as shown in Figure 4.4. The dynamic IFT behaviour was observed, and after 40 minutes, when it reached saturation, the equilibrium IFT values were reported (Figure 4.4) [55]. The difference in the IFT value before it reaches equilibrium was because of the simultaneous adsorption of surfactant molecules at the oil–water interface and also the amount of ionized and un-ionized acid group at the interface [67].

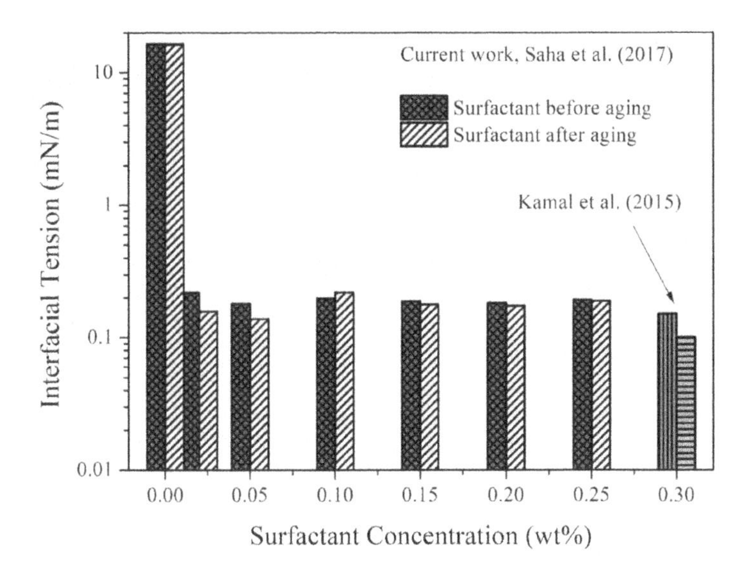

FIGURE 4.4
IFT between crude oil and surfactant solution for both aged and non-aged surfactant using formation water as the aqueous medium.

The equilibrium IFT reduces to 10^{-1} mN/m as the concentration of surfactant increases to 0.02 wt% and remains constant thereafter even after enhancing the concentration to 0.25 wt%. The desirable decrease in the IFT values was observed due to the accumulation of a higher amount of surfactant molecules at the interface. The reduction in the IFT curve flats off at a higher surfactant concentration, which was probably due to saturation of the adsorbed surfactant molecules at the interfacial film [68]. The graph for the aged surfactant samples showed similar behaviour with negligible variation in their IFT values at respective surfactant concentrations. This behaviour directly exposed the thermal stability of the surfactant with respect to IFT analyses. The same behaviour of negligible variation in IFT with aged and non-aged surfactants (named as Surfactant A) was observed by other authors (Kamal et al. [66]) at a higher concentration of 0.3 wt% (Figure 4.4). Therefore, from the collected data, we can conclude that the surfactant is thermally stable.

4.5 Adsorption Isotherms

The adsorption studies were performed by mixing Hapjan rock samples (collected from Assam Oil field, India) with the surfactant solution at different concentrations in the ratio of 1:5 respectively. The mixing was carried out in an orbital shaker at room temperature until equilibrium was achieved. The sample was then centrifuged and analysed by UV-Visible spectroscopy to estimate the adsorption quantity [55]. The experimental data are then fitted in different isotherm models like Freundlich, Langmuir, Linear and Temkin. The data reveal that the Langmuir model was the best fit with a 0.99 regression coefficient and a lowest average relative error of 2.6%. The next appropriate model was Temkin based on R^2 (0.88), but the average relative error was 5.5%. Freundlich isotherm fitting was not much significant with respect to R^2 (0.81) though better than linear isotherm model (R^2=0.46) and the error obtained were 3.5% and 22% respectively. Therefore, based on the fitting, it was observed that the Langmuir isotherm is the perfect isotherm model to describe the surfactant adsorption characteristics. The parameters or constant values for this fitting are found to be q_o (mg/g)=16.89 and K_L (L/mg)=0.0025. Thus, it can be concluded that the adsorbent is homogeneous in nature undergoing monolayer adsorption with no interaction between the adsorbed molecules [45,47,55].

4.5.1 Adsorption Kinetic Models

The adsorption kinetics was performed using Lagergren's pseudo-first-order rate, pseudo-second-order rate, intra-particle diffusion and Elovich models and the plots are represented in Figure 4.5a–d, respectively. The fittings at different surfactant concentrations (500–3500 mg/L) using the above kinetic

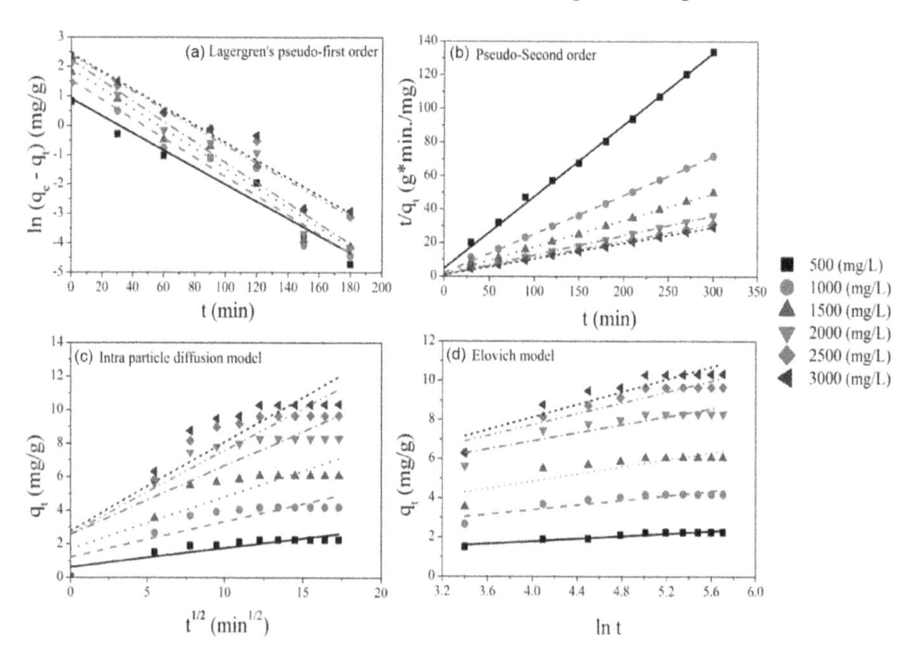

FIGURE 4.5
Adsorption kinetics of Triton X-100 on Hapjan reservoir rock (Assam, India) (a) Lagergren's pseudo-first-order, (b) pseudo-second-order, (c) intraparticle diffusion model and (d) Elovich model.

TABLE 4.1

Adsorption Kinetics Data Obtained at Different Concentrations of Triton X-100

	Lagergren's Pseudo-First Order		Pseudo-Second Order		Intraparticle Diffusion		Elovich	
Surfactant Concentration (mg/L)	R^2	Average Relative Error (%)	R^2	Average Relative Error (%)	R^2	Average Relative Error (%)	R^2	Average Relative Error (%)
500	0.95	9.14	0.997	3.94	0.79	31.46	0.91	8.90
1000	0.94	9.35	0.998	3.43	0.75	32.44	0.81	9.86
1500	0.95	10.58	0.997	3.76	0.75	32.94	0.74	10.97
2000	0.95	9.06	0.998	2.79	0.73	36.62	0.80	9.71
2500	0.95	10.31	0.997	3.92	0.78	37.41	0.88	10.71
3000	0.94	9.44	0.997	4.01	0.78	37.16	0.85	11.77
Average	0.94	9.65	0.997	3.64	0.77	34.67	0.84	10.32

models were executed and the average value of regression coefficient (R^2) and the average value of average relative error were reported as shown in Table 4.1. The average R^2 and the average value of average relative error for Lagergren's pseudo-first-order rate were found to be 0.95% and 9.65%, respectively [55]. The best fitting was obtained for pseudo-second-order, which resulted in an R^2 and an average relative error value of 0.99% and 3.64% respectively (the value reported is the average value). The R^2 data for intra-particle diffusion and Elovich are 0.76 and 0.84, and their corresponding average relative errors are 34.67% and 10.32%, respectively. Similar kinetic studies were performed by several others in order to predict the surfactant adsorption behaviour on the rock surface [39,45].

4.6 Influence of Salinity and Temperature on Adsorption Capacity

The amount of surfactant adsorbed on the rock surface depends largely on the salinity and temperature of the system as depicted in Figure 4.6. Initially, at 30°C, the adsorption quantity was 7.53 mg/g which enhances to 10.87 mg/g as the aqueous system shifted from millipore water to saline (formation water ~4500 ppm). The same experiment was conducted at a higher temperature of 70°C during which similar behaviour of enhancement in adsorption quantity from 4.86 to 6.23 mg/g was observed with the introduction of saline water. This enhancement in the adsorption quantity of surfactant with salinity was because of the compression of the electric double layer present at the adsorbent surface, which reduces the repulsion force between the adsorbent and the adsorbate [39,69].

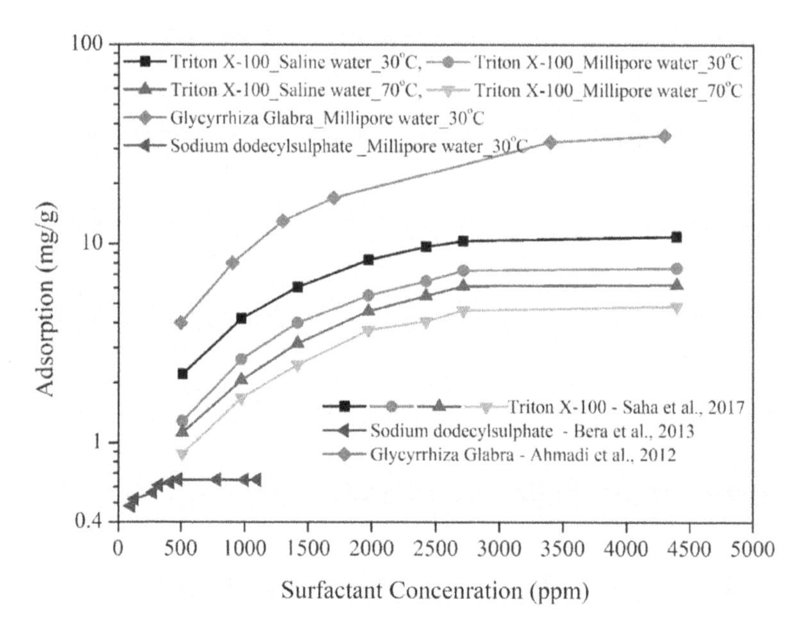

FIGURE 4.6
Adsorption capacity of different surfactants with variation in temperature and formation water (salinity).

The temperature variation showed opposite behaviour in the adsorption quantity of surfactant on the rock surface. At room temperature of 30°C, the adsorption quantity of surfactant was high, which however reduces as the temperature increases. A reduction in the quantity of around 43% was observed as the temperature changes from 30°C to 70°C with millipore water as the base fluid. A similar reduction of around 35.5% was detected as the base fluid shifted from millipore to saline water system. This huge reduction in adsorption quantity was possible due to the temperature effect during which the rate of adsorption of the adsorbate on the surface and in the interior pores of adsorbent reduces drastically because of viscosity reduction. The detailed temperature effect was further analysed by evaluating thermodynamic parameters like enthalpy, entropy and Gibbs free energy [39].

Therefore, the adsorption capacity of surfactant was found to be directly proportional to salinity and inversely proportional to temperature. However, the adsorption capacity of surfactant varies depending on the nature of surfactant and rock surfaces as illustrated in Figure 4.6. The adsorption of novel surfactant (Glycyrrhiza Glabra) on carbonate rock at room temperature was above 20 mg/g before it reaches the equilibrium/saturation point [45]. On the other hand, a lower adsorption of around 0.65 mg/g was reported for a sodium dodecylsulphate (SDS)–sand surface system [39].

4.7 Adsorption Thermodynamic Parameters

The thermodynamic parameters for the adsorption process can be examined by exploring the plot between Langmuir constant and temperature and determining the value of enthalpy and entropy [55]. In this system, the Gibbs free energy derived at different temperatures of 30°C, 40°C, 50°C, 60°C and 70°C was found to be −2.318, −1.757, −1.333, −0.857 and −0.538 kJ/mol, respectively. The negative Gibbs value illustrates that the adsorption system is feasible and spontaneous. The enthalpy obtained was $\Delta H = -15.846$ kJ/mol, which demonstrates the exothermic nature of the process and due to which the adsorption capacity reduces at higher temperature [55]. Similarly, the entropy value of −0.045 kJ/mol k indicates the reduction in molecule randomness at the interface and thus lowers the adsorption quantity at a higher temperature [39].

4.8 Role of Rock Minerals on Adsorption Quantity

The mineralogy of the rock samples (field and synthetic cores) considered in this study was examined by XRD analysis and is already reported in our publication [55]. The different types of rock mineralogy and their percentages were quantified and their corresponding adsorption was identified. The investigation showed that rock with a higher illite mineral undergoes maximum adsorption followed by feldspar > montmorillonite > kaolinite. The plot between mineralogy content and adsorption quantity which is already published [55] revealed the validity of the model with R^2 values of 0.95, 0.92, 0.91 and 0.81 for illite, feldspar, montmorillonite and kaolinite, respectively. The adsorption process depends largely on the above four mineralogy as the R^2 values for those mineralogy were found to be 0.98. Apart from mineralogy, the surface area of the rock samples may also play an important role in adsorption. The natural rock used has a higher surface area of 4.97 m^2/g as compared to Berea cores, which have a surface area of 2.1 m^2/g. The surface area of Berea cores as reported by other authors was further lower as 0.82–1.23 m^2/g [70,71]. It could be due to such surface area that natural reservoir rock showed higher adsorption as compared to synthetic or Berea cores. However, a detailed investigation on the effect of surface area on adsorption characteristics has to be carried out further for its validation.

The adsorption quantity for different rocks and the financial loss encountered are estimated in Table 4.2. The estimation was performed by considering the total reservoir rock volume as $3 \times 10^9 m^3$ with density and porosity as mentioned in Table 4.2. The loss encountered using the surfactant Triton X-100 was found to be almost 0.48 billion $, and this economical loss is on the

TABLE 4.2

Adsorption Capacities and Cost Analysis of Natural and Synthetic Cores at 70°C Using Synthetic Formation Water (4445 ppm Salinity)

S. No	Cores	Core Types	Gas Permeability (mD)	Brine Permeability (mD)	Porosity (%)	Density (kg/m^3)	Adsorption (mg/g)	Triton X-100 Cost (Rs/kg)	Adsorption Loss (Crores Rs)	Adsorption Loss (billion $[a])	Reference
1	Hapjan (Assam)	Natural	20.2	-	20.2	2.45	6.23	3454	3194	0.48	This work
2	Jorajan (Assam)	Natural	2	-	10.4	2.32	8.03	3454	2007	0.30	This work
3	Carbon Tan	Synthetic	42	10–12	12.2–17.7	2.19	4.56	3454	1262–1832	0.19–0.27	This work
4	Idaho Gray	Synthetic	7187–7956	2150–2400	29	1.82	5.07	3454	2772	0.42	This work
5.	Gray Berea	Synthetic	200–315	60–100	19–20	2.11	4.17	3454	1732–1824	0.26–0.27	This work
6	Sandstone	Synthetic	-	-	-	-	1.5	3454	-	-	Muherei et al. [44]

[a] 1$ = 66.85 INR.

higher side as compared to the data reported during the surfactant (DBS)–kaolinite adsorption process [49]. The only reason for the higher cost was the greater rock volume, which is 30 km^2 and 100 m depth with respect to 1 acre and 3 m depth as considered in the surfactant–kaolinite system [49].

4.9 Conclusions

In this chapter, the desired surfactant was selected based on the IFT reduction behaviour between different surfactants and Assam crude oil (India). A total of seven different surfactants were screened and based on the IFT reduction, Triton X-100 was selected. The effect of salinity on the IFT behaviour was also examined. The thermal stability of the chosen surfactant was then explored by TGA, NMR and FTIR studies. Additionally, the impact of rock characteristics on the adsorption capacity was examined. The adsorption data showed Langmuir isotherm and pseudo-second-order kinetics as the best-fitted models with an R^2 value of 0.99 for each. The average relative errors achieved for the Langmuir isotherm and pseudo-second-order kinetics were 2.58% and 3.64%, respectively. The thermodynamic parameters conveyed the spontaneity, feasibility and exothermic behaviour of the adsorption system. Furthermore, the role of rock mineralogy on the adsorption characteristic followed the following order: illite > feldspar > montmorillonite > kaolinite.

References

1. C. D. Yuan, W. F. Pu, X. C. Wang, L. Sun, Y. C. Zhang, and S. Cheng, Effects of interfacial tension, emulsification, and surfactant concentration on oil recovery in surfactant flooding process for high temperature and high salinity reservoirs, *Energy and Fuels*, vol. 29, pp. 6165–6176, 2015.
2. J. J. Sheng, Review of surfactant enhanced oil recovery in carbonate reservoirs, *Advances in Petroleum Exploration and Development*, vol. 6, pp. 1–10, 2013.
3. R. Saha, R. Uppaluri, and P. Tiwari, Effects of interfacial tension, oil layer break time, emulsification and wettability alteration on oil recovery for carbonate reservoirs, *Colloids and Surfaces A*, vol. 559, pp. 92–103, 2018.
4. T. Austad, B. Matre, J. Milter, A. Saevareid, and L. Oyno, Chemical flooding of oil reservoirs 8: Spontaneous oil expulsion from oil- and water-wet low permeable chalk material by imbibition of aqueous surfactant solutions, *Colloids and Surfaces A: Physicochemical and Engineering Aspects*, vol. 137, pp. 117–129, 1998.
5. L. Chen, G. Zhang, J. Ge, P. Jiang, J. Tang, and Y. Liu, Research of the heavy oil displacement mechanism by using alkaline/surfactant flooding system, *Colloids and Surfaces A: Physicochemical and Engineering Aspects*, vol. 434, pp. 63–71, 2013.

6. M. Budhathoki, T.-P. Hsu, P. Lohateeraparp, B. L. Roberts, B.-J. Shiau, and J. H. Harwell, Design of an optimal middle phase microemulsion for ultra high saline brine using hydrophilic lipophilic deviation (HLD) method, *Colloids and Surfaces A: Physicochemical and Engineering Aspects,* vol. 488, pp. 36–45, 2016.
7. Q. Liu, M. Dong, S. Ma, and Y. Tu, Surfactant enhanced alkaline flooding for Western Canadian heavy oil recovery, *Colloids and Surfaces A: Physicochemical and Engineering Aspects,* vol. 293, pp. 63–71, 2007.
8. D. W. Green and G. P. Willhite, *Enhanced Oil Recovery* vol. 6, SPE Textbook Series. SPE: Richardson, TX, 1998.
9. A. O. Gbadamosi, R. Junin, M. A. Manan, A. Agi, and A. S. Yusuff, An overview of chemical enhanced oil recovery: Recent advances and prospects, *International Nano Letters,* vol. 9, pp. 171–202, 2019.
10. M. S. Kamal, I. A. Hussein, and A. S. Sultan, Review on surfactant flooding: Phase behavior, retention, IFT and field applications, *Energy and Fuels,* vol. 31, pp. 7701–7720, 2017.
11. J. J. Sheng, Comparison of the effects of wettability alteration and IFT reduction on oil recovery in carbonate reservoirs, *Asia-Pacific Journal of Chemical Engineering,* vol. 8, pp. 154–161, 2013. doi: 10.1002/apj.1640.
12. J. J. Sheng, Status of surfactant EOR technology, *Petroleum,* vol. 1, pp. 97–105, 2015. doi: 10.1016/j.petlm.2015.07.003.
13. J. M. Maerker and W. W. Gale, Surfactant flood process design for Loudon, *SPE Reservoir Engineering,* vol. 7, pp. 36–44, 1992.
14. W. W. Gale and E. I. Sandvik, Tertiary surfactant flooding: Petroleum sulfonate composition-efficacy studies, *Society of Petroleum Engineers Journal,* vol. 13, pp. 191–199, 1973.
15. D. C. Standnes and T. Austad, Wettability alteration in chalk 2: Mechanism for wettability alteration from oil-wet to water-wet using surfactants, *Journal of Petroleum Science & Engineering,* vol. 28, pp. 123–143, 2000.
16. K. Jarrahian, O. Seiedi, M. S. Sheykhan, M. V, and S. Ayatollahi, Wettability alteration of carbonate rocks by surfactants: A mechanistic study, *Colloids and Surfaces A: Physicochemical and Engineering Aspects,* vol. 410, pp. 1–10, 2012.
17. A. Bera and H. Belhaj, Ionic liquids as alternatives of surfactants in enhanced oil recovery: A state-of-the-art review, *Journal of Molecular Liquids,* vol. 224, pp. 177–188, 2016.
18. T. F. Tadros, *An Introduction to Surfactants.* Walter de Gruyter GmbH, Berlin/Boston, MA, 2014.
19. A. A. Olajire, Review of ASP EOR (alkaline surfactant polymer enhanced oil recovery) technology in the petroleum industry: Prospects and challenges, *Energy,* vol. 77, pp. 963–982, 2014.
20. M. A. Bezerra, M. A. Z. Arruda, and S. L. C. Ferreira, Cloud point extraction as a procedure of separation and pre-concentration for metal determination using spectroanalytical techniques: A review, *Applied Spectroscopy Reviews,* vol. 40, pp. 269–299, 2005.
21. A. Belhaj, K. A. Elraies, S. M. Mahmood, N. N. Zulkifi, S. Akbari, and O. S. Hussien, The effect of surfactant concentration, salinity, temperature, and pH on surfactant adsorption for chemical enhanced oil recovery: A review, *Journal of Petroleum Exploration and Production Technology,* vol. 10, pp. 125–137, 2020.
22. M. S. Kamal, A review of Gemini surfactants: Potential application in enhanced oil recovery, *Journal of Surfactants and Detergents,* vol. 19, pp. 223–236, 2016.

23. R. Saha, R. Uppaluri, and P. Tiwari, Impact of natural surfactant (Reetha), polymer (Xanthan Gum), and silica nanoparticles to enhance heavy crude oil recovery, *Energy and Fuels*, vol. 33, pp. 4225–4236, 2019.
24. C. Schmitt, B. Grassl, G. Lespes, J. Desbrieres, V. Pellerin, and S. Reynaud, Saponins: A renewable and biodegradable surfactant from its microwave-assisted extraction to the synthesis of monodisperse lattices, *Biomacromolecules*, vol. 15, pp. 856–862, 2014.
25. S. Zendehboudi, M. A. Ahmadi, A. R. Rajabzadeh, N. Mahinpey, and I. Chatzis, Experimental study on adsorption of a new surfactant onto carbonate reservoir samples-application to EOR, *Canadian Journal of chemical Engineering*, vol. 91, pp. 1439–1449, 2013.
26. M. A. Ahmadi and S. Shadizadeh, Experimental and theoretical study of a new plant derived surfactant adsorption on quartz surface: Kinetic and isotherm methods, *Journal of Dispersion Science and Technology*, vol. 36, pp. 441–452, 2015.
27. M. A. Ahmadi and S. R. Shadizadeh, Experimental investigation of a natural surfactant adsorption on shale-sandstone reservoir rocks: Static and dynamic conditions, *Fuel*, vol. 159, pp. 15–26, 2015.
28. P. Raffa, A. A. Broekhuis, and F. Picchioni, Polymeric surfactants for enhanced oil recovery: A review, *Journal of Petroleum Science and Engineering*, vol. 145, pp. 723–733, 2016.
29. L. Chen, G. Zhang, J. Ge, P. Jiang, J. Tang, and Y. Liu, Research of the heavy oil displacement mechanism by using alkaline/surfactant flooding system, *Colloids and Surfaces A: Physicochemical and Engineering Aspects*, vol. 434, pp. 63–71, 2013.
30. A. Samanta, K. Ojha, A. Sarkar, and A. Mandal, Surfactant and surfactant-polymer flooding for enhanced oil recovery, *Advances in Petroleum Exploration and Development*, vol. 2, pp. 13–18, 2011.
31. L. Fu, G. Zhang, J. Ge, K. Liao, H. Pei, P. Jiang, et al., Study on organic alkali-surfactant-polymer flooding for enhanced ordinary heavy oil recovery, *Colloids and Surfaces A: Physicochemical and Engineering Aspects*, vol. 508, pp. 230–239, 2016.
32. D. Wang, J. Cheng, J. Wu, Z. Yang, Y. Yao, and H. Li, Summary of ASP pilots in Daqing oil field, in *SPE Asia Pacific Improved Oil Recovery Conference*, 25–26 October, Kuala Lumpur, Malaysia, 1999.
33. M. A. Ahmadi, M. Galedarzadeh, and S. R. Shadizadeh, Wettability alteration in carbonate rocks by implementing new derived natural surfactant: Enhanced oil recovery applications, *Transport in Porous Media*, vol. 106, pp. 645–667, 2015.
34. M. A. Ahmadi and S. R. Shadizadeh, Nano-surfactant flooding in carbonate reservoirs: A mechanistic study, *European Physical Journal Plus*, vol. 132, pp. 1–13, 2017.
35. F. D. S. Curbelo, V. C. Santanna, E. L. B. Neto, T. V. Dutra, T. N. C. Dantas, A. A. D. Neto, et al., Adsorption of nonionic surfactants in sandstones, *Colloids and Surfaces A: Physicochemical and Engineering Aspects*, vol. 293, pp. 1–4, 2007.
36. M. A. Ahmadi and S. R. Shadizadeh, Adsorption of a nonionic surfactant onto a silica surface, *Energy Sources, Part A: Recovery, Utilization, and Environmental Effects*, vol. 38, pp. 1455–1460, 2016.
37. M. A. Ahmadi and S. R. Shadizadeh, Experimental investigation of adsorption of a new nonionic surfactant on carbonate minerals, *Fuel*, vol. 104, pp. 462–467, 2013.

38. P. Somasundaran and L. Zhang, Adsorption of surfactants on minerals for wettability control in improved oil recovery processes, *Journal of Petroleum Science & Engineering,* vol. 52, pp. 198–212, 2006.
39. A. Bera, T. Kumar, K. Ojha, and A. Mandal, Adsorption of surfactants on sand surface in enhanced oil recovery: Isotherms, kinetics and thermodynamic studies, *Applied Surface Science,* vol. 284, pp. 87–99, 2013.
40. H. ShamsiJazeyi, R. Verduzco, and G. J. Hirasaki, Reducing adsorption of anionic surfactant for enhanced oil recovery: Part II. Applied aspects, *Colloids and Surfaces A: Physicochemical and Engineering Aspects,* vol. 453, pp. 168–175, 2014.
41. Q. Liu, M. Dong, W. Zhou, M. Ayub, Y. P. Zhang, and S. Huang, Improved oil recovery by adsorption–desorption in chemical flooding, *Journal of Petroleum Science & Engineering,* vol. 43, pp. 75–86, 2004.
42. T. Amirianshoja, R. Junin, A. K. Idris, and O. Rahmani, A comparative study of surfactant adsorption by clay minerals, *Journal of Petroleum Science & Engineering,* vol. 101, pp. 21–27, 2013.
43. C. Wang, X.-L. Cao, L.-L. Guo, Z.-C. Xu, L. Zhang, Q.-T. Gong, et al., Effect of adsorption of catanionic surfactant mixtures on wettability of quartz surface, *Colloids and Surfaces A: Physicochemical and Engineering Aspects,* vol. 509, pp. 564–573, 2016.
44. M. A. Muherei, R. Junin, and A. B. B. Merdhah, Adsorption of sodium dodecyl sulfate, triton X100 and their mixtures to shale and sandstone: A comparative study, *Journal of Petroleum Science & Engineering,* vol. 67, pp. 149–154, 2009.
45. M. A. Ahmadi, S. Zendehboudi, A. Shafiei, and L. James, Nonionic surfactant for enhanced oil recovery from carbonates: Adsorption kinetics and equilibrium, *Industrial and Engineering Chemistry Research,* vol. 51, pp. 9894–9905, 2012.
46. A. Barati-Harooni, A. Najafi-Marghmaleki, A. Tatar, and A. H. Mohammadi, Experimental and modeling studies on adsorption of a nonionic surfactant on sandstoneminerals in enhanced oil recovery process with surfactant flooding, *Journal of Molecular Liquids,* vol. 220, pp. 1022–1032, 2016.
47. A. Barati, A. Najafi, A. Daryasafar, P. Nadali, and H. Moslehi, Adsorption of a new nonionic surfactant on carbonate minerals in enhanced oil recovery: Experimental and modeling study, *Chemical Engineering Research and Design,* vol. 105, pp. 55–63, 2016.
48. S. B. Gogoi, Adsorption of non-petroleum base surfactant on reservoir rock, *Current Science,* vol. 97, pp. 1059–1063, 2009.
49. S. Park, E. S. Lee, and W. R. W. Sulaiman, Adsorption behaviors of surfactants for chemical flooding in enhanced oil recovery, *Journal of Industrial and Engineering Chemistry,* vol. 21, pp. 1239–1245, 2015.
50. W. Kwok, R. E. Hayes, and H. A. Nasr-El-Din, Modelling dynamic adsorption of an anionic surfactant on berea sandstone with radial flow, *Chemical Engineering Science,* vol. 50, pp. 769–783, 1995.
51. A. M. Blokhus, H. HØiland, M. I. Gjerde, and E. K. Ersland, Adsorption of sodium dodecyl sulfate on kaolin from different alcohol–water mixtures, *Journal of Colloid and Interface Science,* vol. 179, pp. 625–627, 1996.
52. J. Novosad, Surfactant retention in berea sandstone-effects of phase behavior and temperature, *Society of Petroleum Engineers Journal,* vol. 22. pp. 962–970, 1982.

53. M. Budhathoki, S. H. R. Barnee, B.-J. Shiau, and J. H. Harwella, Improved oil recovery by reducing surfactant adsorption with polyelectrolyte in high saline brine, *Colloids and Surfaces A: Physicochemical and Engineering Aspects,* vol. 498, pp. 66–73, 2016.
54. M. A. Ahmadi and S. R. Shadizadeh, Adsorption of novel nonionic surfactant and particles mixture in carbonates: Enhanced oil recovery implication, *Energy Fuels,* vol. 26, pp. 4655–4663, 2012.
55. R. Saha, V. S. R. Uppaluri, and P. Tiwari, Effect of mineralogy on the adsorption characteristics of surfactant: Reservoir rock system, *Colloids and Surfaces A: Physicochemical and Engineering Aspects,* vol. 531, pp. 121–132, 2017.
56. H.-R. Li, Z.-Q. Li, X.-W. Song, C.-B. Li, L.-L. Guo, L. Zhang, et al., Effect of organic alkalis on interfacial tensions of surfactant/polymer solutions against hydrocarbons, *Energy Fuels,* vol. 29, pp. 459–466, 2015.
57. L. Zhang, L. Luo, S. Zhao, and J. Y. Yu, *Ultra Low Interfacial Tension and Interfacial Dilational Properties Related to Enhanced Oil Recovery.* Nova Science Publishers, New York, 2008.
58. J. R. Kanicky, J.-C. Lopez-Montilla, S. Pandey, and D. O. Shah, Surface chemistry in the petroleum industry, in Holmberg, K., (Ed.), *Handbook of Applied Surface and Colloid Chemistry.* John Wiley & Sons, Ltd, Gainesville, FL, pp. 251–267, 2001.
59. J. Rudin and D. T. Wasan, Mechanisms for lowering of interfacial tension in alkali/acidic oil systems: Effect of added surfactant, *Industrial & Engineering Chemistry Research,* vol. 31, pp. 1899–1906, 1992.
60. J. Rudin, C. Bernard, and D. T. Wasan, Effect of added surfactant on interfacial-tension and spontaneous emulsification in alkali/acidic oil systems, *Industrial & Engineering Chemistry Research,* vol. 33, pp. 1150–1158, 1994.
61. J. E. Puig, M. T. Mares, W. G. Miller, and E. I. Franses, Mechanism of ultralow interfacial tensions in dilute surfactant-oil-brine systems, *Colloids and Surface A,* vol. 16, pp. 139–152, 1985.
62. A. Bera, A. Mandal, and B. B. Guha, Synergistic effect of surfactant and salt mixture on interfacial tension reduction between crude oil and water in enhanced oil recovery, *Journal of Chemical & Engineering Data,* vol. 59, pp. 89–96, 2014.
63. L. Zhang, L. Luo, S. Zhao, Z. Xu, J. An, and J. Yu, Effect of different acidic fractions in crude oil on dynamic interfacial tensions in surfactant/alkali/model oil systems, *Journal of Petroleum Science and Engineering,* vol. 41, pp. 189–198, 2004.
64. Z. Zhao, C. Bi, Z. Li, W. Qiao, and L. Cheng, Interfacial tension between crude oil and decylmethylnaphthalene sulfonate surfactant alkali-free flooding systems, *Colloids and Surfaces A: Physicochemical and Engineering Aspects,* vol. 276, pp. 186–191, 2006.
65. L. Zhang, L. Luo, S. Zhao, and J. Yu, Studies of synergism/antagonism for lowering dynamic interfacial tensions in surfactant/alkali/acidic oil systems, part 2: Synergism/antagonism in binary surfactant mixtures, *Journal of Colloid Interface Science,* vol. 251, pp. 166–171, 2002.
66. M. S. Kamal, A. S. Sultan, and I. A. Hussein, Screening of amphoteric and anionic surfactants for cEOR applications using a novel approach, *Colloids and Surfaces A: Physicochemical and Engineering Aspects,* vol. 476, pp. 17–23, 2015.
67. Z. Zhao, Z. Li, W. Qiao, and L. Cheng, Dynamic interfacial tension between crude oil and dodecyl methylnaphthalene sulfonate surfactant flooding systems, *Energy Sources, Part A,* vol. 29, pp. 207–215, 2007.

68. C. A. Miller and K. H. Raney, Solubilization-emulsification mechanisms of detergency, *Colloids and Surfaces A: Physicochemical and Engineering Aspects,* vol. 74, pp. 169–175, 1993.
69. A. V. Pethkar and K. M. Paknikar, Recovery of gold from solutions using cladosporium cladosporioides biomass beads, *Journal of Biotechnology,* vol. 63, pp. 121–136, 1998.
70. E. C. Donaldson, R. F. Kendall, B. A. Baker, and F. S. Manning, Surface-area measurement of geologic materials, *Society of Petroleum Engineers Journal,* vol. 15, pp. 111–116, 1975.
71. P. L. Ghurcher, P. R. French, J. G. Shaw, and L. L. Schramm, Rock properties of berea sandstone, baker dolomite, and Indiana limestone, *Society of Petroleum Engineers Journal,* pp. 431–446, 1991. Paper presented at the SPE International Symposium on Oilfield Chemistry, Anaheim, California, February 1991, Paper Number: SPE-21044-MS, doi: 10.2118/21044-MS.

5

Nanofluid Flooding for Oil Recovery

5.1 Introduction to Nanofluid Flooding

Nanofluids are those fluids in which nanoparticles are suspended in a base fluid. Nanoparticles are fine solid particles of 0.1–100 nm size and are usually found in a superfine powder form. Nanoparticles are greatly used in several industrial applications like food, textile, biology, medicine, electronics, chemistry, oil and gas sector and so on [1–9]. The demand for nanotechnology is tremendous in the current scenario and it all started after 1984 when the first nanomaterial was discovered.

In the petroleum industry, the recovery of heavy crude oil from reservoirs is challenging because of the crude oil properties and therefore in most scenarios, polymer flooding is preferred [3,10–13]. The process of polymer flooding was initiated in the early 1950s and has evolved as a promising method for the tertiary recovery of heavy crude oil [10,12,14–17]. Polymer solution has desired rheological properties, which drastically reduce the viscous fingering and promote the oil recovery factor. Additionally, polymer can emulsify the crude oil to form stable oil–water emulsion due to its surface-active properties [18–21]. The success and performance of polymer flooding are dependent on the temperature and salinity of the system [22–25]. The enhancement in salinity of the aqueous phase can degrade the polymer slug, thus affecting the oil recovery and therefore in such a scenario, preflush slug before injection of polymer slug is preferable to reduce the detrimental effect of salinity [13]. The degradation of polymer slug is basically due to the presence of an amide group, which reacts to form a carboxylic compound [26,27]. The temperature in a similar manner can significantly hamper the rheological properties thus reducing the sweep efficiency and resulting in a lower oil recovery [10,28,29]. Moreover, the enhancement in temperature can increase the rate of oil droplet coalescence, which affects emulsion stability hindering the oil recovery factor [30,31].

Considering the limitation of chemical flooding for enhanced oil recovery (EOR), nanofluid flooding was introduced, which showed positive results in terms of higher oil recovery [29,32–36]. Nanofluid prepared from surfactant–polymer–nanoparticles mixtures showed successful improvement in oil recovery achieved by reducing the interfacial tension (IFT), improving rheological properties and most importantly reducing the droplet coalescence [29,36]. Nanoparticles, due to their high surface area, undergo adsorption on the oil droplets and therefore reduce coalescence of droplets improving emulsion stability [37]. Similarly, with a nanoparticles–polymer system, improvement in rheological properties can be accomplished, which introduces desirable displacement efficiency as required for higher oil recovery [29,34,38,39]. The synergy of polymer and nanoparticles introduces bridging-induced flocculation, which assists in increasing the rheological activity [40–42]. Moreover, the synergy behaviour between polymer and nanoparticles can knock out the limitation of high temperature and high salinity as encountered during chemical EOR [42]. Some of the examples of nanoparticles employed with polymer flooding are TiO_2, Al_2O_3 and SiO_2 [43,44]. Nanoparticles additionally can further change the wettability of the system from oil-wet to desirable water-wet by undergoing distribution in the pores and throats of the rock [45,46].

The state-of-the-art for nanofluid flooding revealed that minimum investigation on the mechanisms like IFT, emulsion formation and stability, wettability alteration and rheology properties is available. The studies on chemical EOR is established to some extent; however, deep knowledge on nanofluid is missing. The role of emulsion mechanisms for chemical EOR is available [18,19,21], whereas the involvement of nanoparticles in the same system is not investigated in detail. The relationships between IFT and emulsification processes involving nanoparticles are not executed or explored. The studies on enhancing the emulsion stability at high temperature and salinity were performed to a limited extent [30]. Furthermore, the clog effect of nanoparticles in the pore of the reservoir rock reducing porosity and permeability is in the developing stage.

Therefore, considering the limitation and challenges, a polymer–nanoparticles system has been formulated to perform nanofluid flooding. In this chapter, the stability of nanoparticles by zeta potential and dynamic light scattering, emulsion stability (droplet size), creaming rate, rheological properties, alteration in wettability and nanofluid flooding will be discussed in detail to reduce the state-of-the-art literation gap.

5.2 Methods to Evaluate Nanofluid Stability

The stability of nanoparticles is one of the major criteria to execute successful nanofluid flooding. Nanoparticles, when dispersed in the base fluid, tend to agglomerate, which impacts its stability and hence reduces the essential characteristics [9,47]. The article available on nanofluid illustrated that the dispersion of nanoparticles in aqueous solutions can be classified into three stages. The first stage includes wetting of the nanoparticles followed by demolition of agglomerates and the final step depends on the formation of smaller agglomerates [9,47,48]. The stability of nanoparticles can be understood by the Brownian motion of nanoparticles. The forces of attraction and repulsion between the nanoparticles monitor the stability. For the particles to be in a stable state, the repulsion force should be higher than the attractive force or else, the particle will aggregate into soft and hard agglomerates [47,49]. The repulsive forces can be further calculated by zeta potential values and a higher value of zeta potential indicates strong repulsive forces resulting in better stability of nanoparticles [9,47,50–52]. There are several methods available to detect the stability of nanoparticles and some of them include visual observation (sedimentation), particle size analyser and zeta potential.

Figure 5.1 represents a schematic diagram of the behaviour of nanoparticles in a chemical/polymer and reservoir saline water system. The time-dependent visual observations were monitored for up to 10 days to observe the sedimentation phenomena. The study depicts that for a xanthan gum (XG) polymer solution, the silica nanoparticles (SNPs) were stable for 10 days. However, for the same system, severe sedimentation behaviour (above 70%) was noticed within 24 hours when the polymer solution was replaced by reservoir saline water and complete settlement of particles within 10 days [28].

Visual observation may not be the perfect method to determine the stability of nanoparticles in nanofluid solutions. Therefore, the size distribution of particles in the polymer solutions and their corresponding zeta potential data were identified to confirm the stability of the polymer–nanoparticle/nanofluid system. The average size of the cluster nanoparticles and their variation at different nanoparticle concentrations are shown in Table 5.1. The behaviour of the clusters or agglomeration was further monitored for 10 days to forecast the stability of the system. The average sizes of the nanoparticle clusters were enhanced

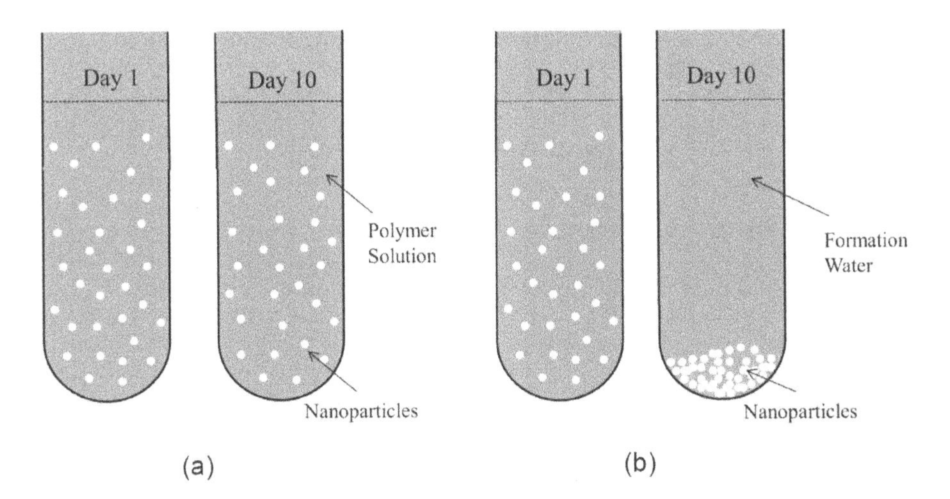

FIGURE 5.1
Schematic diagram illustrating time-dependent nanoparticles behaviour in a polymer solution and saline water system: (a) suspension and (b) sedimentation [28].

TABLE 5.1

Size Distribution of Nanoparticles Suspended in Polymer Solutions (Xanthan Gum) of 5000 ppm at Temperature 30°C

Sr. No.	Polymer Concentration (ppm)	Nanoparticles Concentration (wt%)	Diameter (μm)	
			Day 1	Day 10
1	5000	0.1	3.95	3.73
2	5000	0.3	4.49	4.18
3	5000	0.3	4.64	4.03

as the concentration of nanoparticles increased from 0.1% to 0.5 wt%. This variation in the sizes of the clusters was found to be around ~3500 to 5000 nm. However, the data obtained showed a negligible reduction in the agglomeration sizes of the nanoparticles after 10 days for all cases (Table 5.1). This difference in the agglomeration size with duration was observed because of the settled behaviour of the large agglomerates with respect to the suspended small agglomerate nanoparticles [29]. Hence, such insignificant variation in the average size of cluster nanoparticles specifies negligible sedimentation and thus confirms the stability of the nanofluid solution. The zeta potential values for the same system (5000 ppm xanthan gum and 0.1–0.5 wt% silica nanoparticles) are depicted in Figure 5.2. The value obtained at different nanoparticle concentrations was within the range of −21 to −27 mV with a deviation of around <±2%. The zeta value after a period of 10 days showed minimum differences in the range of ~2% to 4%. Hence, from this observed value of zeta potential with duration, we can conclude that the system is stable. The system is considered to be stable only when the zeta potential is in the range of ±30 mV but in this case, such value

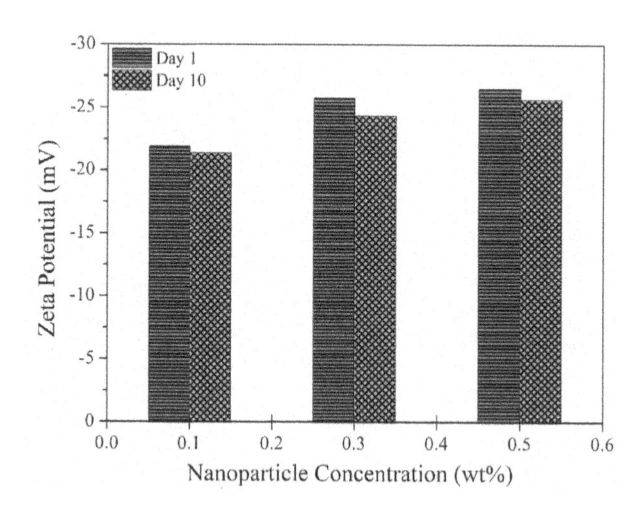

FIGURE 5.2
Stability analysis of nanoparticles in 5000 ppm xanthan gum polymer solution by zeta potential value at a temperature of 30°C.

could not be reached because of the presence of ions in the used formation water/aqueous phase. A detailed study on the effects of ions towards the zeta potential value is also available [53]. In this study, an optimum concentration of 5000 ppm polymer was considered based on the emulsification investigation as explained in Section 5.5.

5.3 Influence of Nanofluid on Rheological Properties

The theoretical estimation of the viscosity for solutions possessing different amounts of suspended particles can be predicted by the Einstein formula [9,54]. The viscosity of nanofluid solutions is expected to be high because of the presence of suspended particles in the base fluid [3,9,29]. Hence, the viscosity distributions of nanofluid (polymer–nanoparticles) solutions considered in our study are discussed [28]. The viscosity of the polymer solution (without nanoparticles) was observed to be 2460 mPa s at room temperature (30°C), which jumps to 2970 mPa s as the concentration of nanoparticles reaches 0.5 wt%. A similar development in viscosity (from 593 to 771 mPa s) for the nanofluid solutions was spotted as the temperature rose to 80°C. The desired properties of improved viscosity were achieved because of the establishment of spatial macromolecular network structures assisted by the presence of nanoparticles [42,55,56]. A drastic reduction in the viscosities was witnessed at a high temperature, which is due to the destruction of this complex network encountered due to the weak bonding interaction [42,57]. Though high temperature reduces the viscosity, the presence of nanoparticles in the nanofluid solutions showed better performance with respect to its absence. The available research performed by various authors concluded that the viscosity of nanofluid solution depends on parameters like the types of nanoparticles used, their concentration and properties of the base fluid [9,58–61]. The frequency sweep study was also executed to observe the storage and loss modulus behaviour. Initially, at a lower frequency, the storage modulus was dominant whereas at a higher frequency, the solution showed viscoelastic properties [36]. The data revealed no crossover point, which clearly indicates the gel activity of the solution in the bulk phase [28,62,63].

5.4 Influence of Nanofluid on Interfacial Tension

The reduction in IFT between nanofluid/nanoparticles–polymer solutions and crude oil at different temperatures is represented in Figure 5.3. The IFT encountered at the beginning (i.e. in the absence of nanoparticles and at a

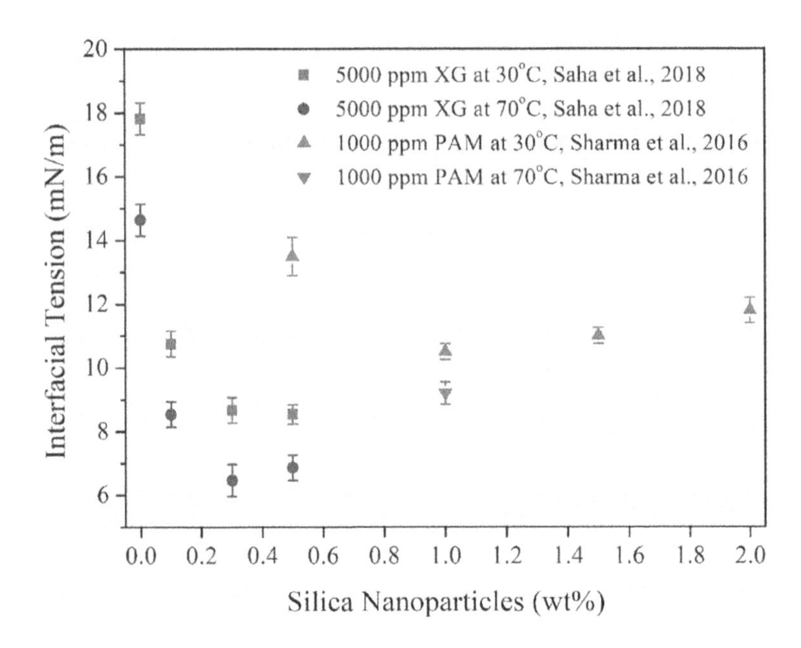

FIGURE 5.3
IFT behaviour between crude oil and nanofluid/nanoparticles–polymer solutions at different nanoparticle concentrations and temperatures.

room temperature of 30°C) was measured to be 17.8 mN/m. The IFT of the system at the same room temperature then approaches a minimum value of 8.54 mN/m as the concentration of nanoparticles in the nanofluid solution reaches 0.5 wt%. As the temperature reaches 70°C, the IFT reduces to the lowest value of 6.86 mN/m at a concentration of 0.5 wt% nanoparticles from the initial IFT of 14.6 mN/m (no nanoparticles). The reduction in IFT occurred because of numerous factors like enhanced surface–volume ratio, greater adsorption, hydrophilic–hydrophobic interaction and suspension stability [64,65]. However, the IFT data obtained showed negligible variation in IFT reduction as the concentration of silica nanoparticles touches ≥0.3 wt% at both 30°C and 70°C. Therefore, silica nanoparticles concentration of 0.3 wt% was chosen as optimum based on the limitation of IFT reduction. The inability of nanoparticles to reduce IFT beyond its optimum value was because of the saturation of nanoparticles at the interface [29,65]. Comparable studies using Indian crude were investigated in which the concentration of silica nanoparticles was varied from 1 to 2 wt% with the temperature at 30°C and 70°C [29]. The data obtained are also similar except for the range of silica nanoparticles (0.5–2 wt% nanoparticles), which depends on the crude oil and nanofluid interaction properties. The IFT studies using nanofluid are useful as it governs the flow of residual oil in the pores/pore throat of the reservoir rock [66,67].

5.5 Effect of Nanofluid on Emulsion Properties

5.5.1 Nanofluid for Emulsion Stability

The xanthan gum polymer solution has the capability to emulsify the crude oil when mixed together as depicted in Figure 5.4. The study illustrates the formation of emulsion and its stability with the concentration of xanthan gum varied from 100 to 500 ppm. The coverage of xanthan gum on the crude oil droplets decides the stability factor of the emulsion. Initially, at 100 ppm, the poor coverage of xanthan gum over the oil droplets causes droplets coalescence and this coalescence severely affects the emulsion stability [68,69]. However, as the concentration of xanthan gum enhances to 5000 ppm, the coverage is improved, which reduces the coalescence effect by developing bridging between droplets and supports flocculation [70]. Therefore, the emulsion achieved with 5000 ppm xanthan gum solutions was stable and hence such concentration is fixed as optimum (Figure 5.4).

After fixing the concentration of polymer solution, silica nanoparticles in the range of 0.1–0.5 wt% were added to the system to evaluate the stability of the emulsion. The visual observation and microscopic images of the samples were analyzed to perform the emulsion stability behaviour [28]. The visual observation showed the enhancement in emulsion stability from 15 days (without nanoparticles) to more than 25 days (with nanoparticles). The microscopic images reveal the droplet size formation and their distribution as the concentration of nanoparticles increases from 0 to 0.5 wt%. The droplet size distribution in the absence of nanoparticles varied in the range of 0–22 μm, which reduces to 0–13 μm at 0.5 wt% nanoparticles (Figure 5.5). The average droplet diameters obtained were more or less the same for 0.3 and 0.5 wt% nanoparticles. This behaviour could have been due to the

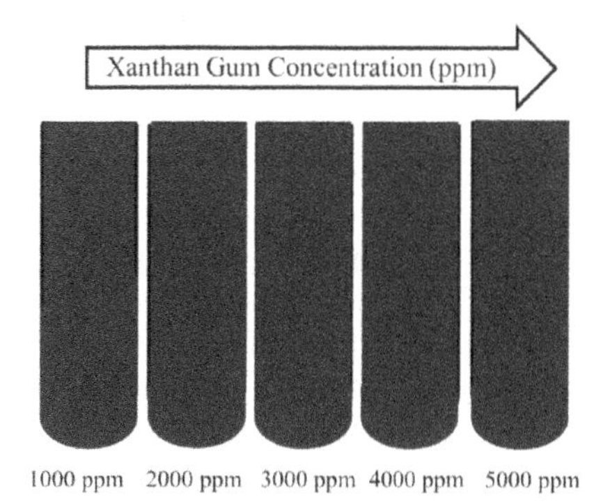

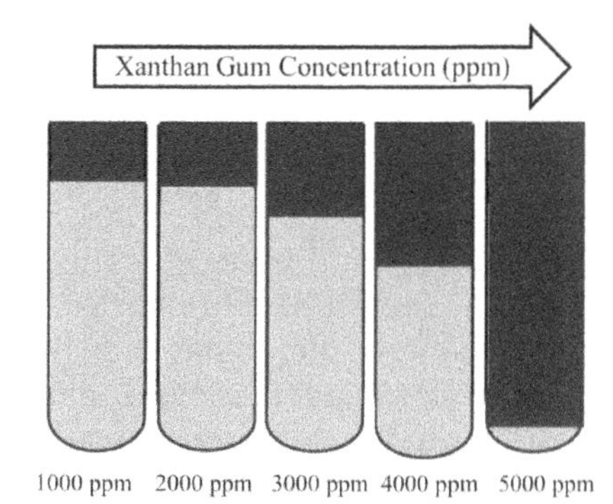

FIGURE 5.4
The emulsion behaviour using crude oil and xanthan polymer solutions at a room temperature of 30°C.

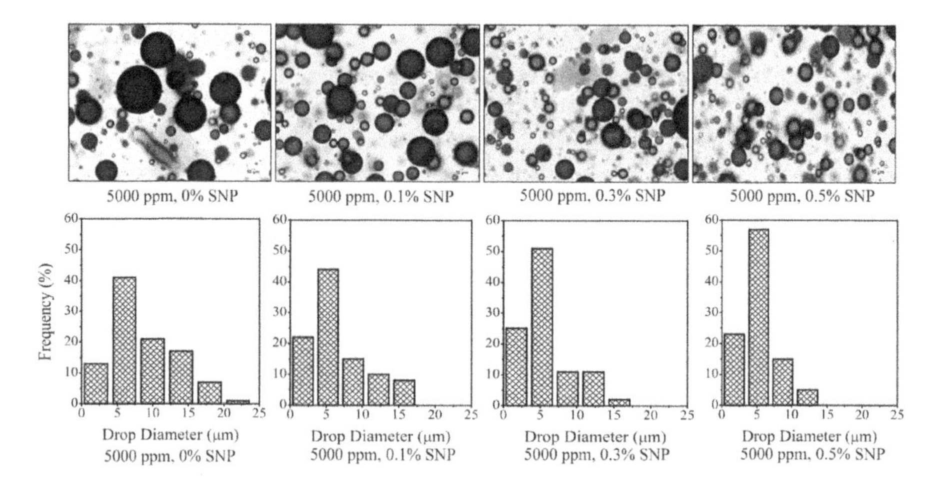

FIGURE 5.5
The microscopic images of the emulsion showing distribution of droplet size at a room temperature of 30°C.

saturation of adsorbed nanoparticles at the interface. Thus, the formation of these small average droplet diameters due to the existence of nanoparticles thereby stabilizes the emulsion by neglecting the coalescence induced due to reduction in cohesive forces between droplets. Hence, based on the emulsion behaviour, IFT reduction and cost of the chemicals, an optimum formulation of 5000ppm xanthan gum and 0.3wt% silica nanoparticles were considered. The effect of nanoparticles on emulsion stability has also been investigated by several other authors [30,36,71]. The authors reported that the coalescence rate for droplets is monitored by the adsorption of nanoparticles on the surface of the droplets [71]. The retention of nanoparticles on the oil droplet was investigated by FESEM characterization and its schematic diagram is represented in Figure 5.9.

5.5.2 Nanofluid for Creaming Index

Creaming is the process of emulsion separation during which the layers of oil and aqueous phase are parted. Therefore to have a detailed understanding of the emulsion behaviour, the creaming rate of the emulsion formed using crude oil and nanofluid solutions was observed and its corresponding creaming index was reported as shown in Table 5.2. The concentrations of nanoparticles in the nanofluid solutions were varied from 0.1 to 0.5wt%. The emulsions formed were put to an uninterrupted state for a particular time (up to 10days) and the corresponding separation was measured at temperatures of 30°C and 80°C. The creaming rate at a room temperature of 30°C was negligible as it showed 79% and 86% creaming index values after 10days for the samples which included 0 and 0.1wt% nanoparticles, respectively. The same

TABLE 5.2

Effect of Nanoparticles on the Creaming Rate for the Emulsion Obtained with Crude Oil–Polymer-Nanoparticles at Different Temperatures

Xanthan Gum (ppm)	Silica Nanoparticles (wt%)	Creaming Index (%) at 30°C			Creaming Index (%) at 80°C		
		Initial Time	Day 1	Day 10	Initial Time	Day 1	Day 10
5000	0	100	89	79	100	8	0
	0.1	100	93	86	100	36	15
	0.3	100	100	100	100	58	15
	0.5	100	100	100	100	58	15

system in which the concentration of nanoparticles was above ≥0.3 wt% does not specify any creaming behaviour and thereby produces a 100% creaming index. However, the behaviour at 80°C changes drastically and it undergoes extreme creaming on the first day itself. The samples with a greater concentration of nanoparticles showed a greater creaming index value but the value was limited beyond 0.3 wt% on day 1. The creaming rate further enhanced with a duration of 10 days and showed a 15% creaming index value with nanoparticles and 100 creaming or 0% creaming index value with the sample that includes no nanoparticles. This revealed the performance of nanoparticles at a higher temperature with respect to their absence. The creaming rate or stability of emulsion depends on the average droplet size (less size better stability) and is greatly monitored by the amount of nanoparticles adsorbed at the interface forming a dense pack [71].

5.5.3 Nanofluid for Emulsion Viscosity

The viscosity of each phase in the system is important as it monitors the overall displacement efficiency, which controls the crude oil recovery. It was observed that improvement in the viscosity favours the diversion of the aqueous phase through the reservoir covering a larger area, which enhances the recovery factor [72,73]. The viscosity inside the reservoir differs due to temperature variation and hence to overcome such effect, viscosity was measured at different temperatures as shown in Figure 5.6. Initially, at a room temperature of 30°C, the viscosity of the nanofluid solutions, as well as emulsions, increases with an increase in the concentration of nanoparticles. Similar behaviour was observed at a higher temperature of 80°C; however, the value of viscosities was less as compared to room temperature. Irrespective of the temperature effect, it was observed that emulsion possess higher viscosity which helps in diverting the aqueous/nanofluid phase towards an untouched region of the oil reservoir thereby increasing the displacement efficiency [74]. The enhancement in the recovery of crude oil with respect to rheological properties (favourable mobility/

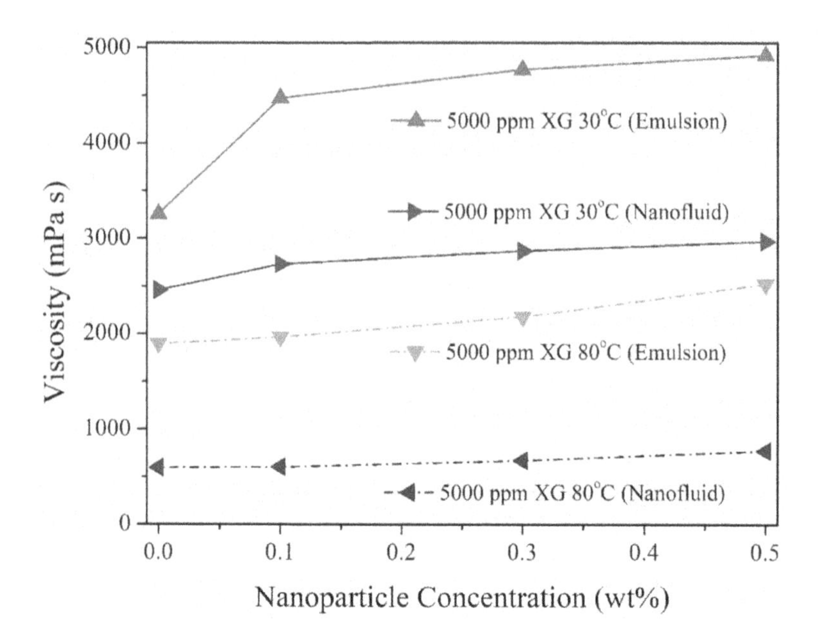

FIGURE 5.6
The effect of temperature on the viscosities of aqueous nanofluid solutions and the emulsions formed at different concentrations of nanoparticles.

displacement efficiency) as explained can be further validated by performing the core flooding experiments at room temperature and at a higher temperature of 80°C.

5.6 Influence of Nanofluid on Wettability Alteration

The changes in the wettability of the system can either reduce the capillary force or increase the capillary number, thereby improveing oil mobility resulting in improved oil recovery. The alteration in the wettability of the reservoir rock using nanofluid or nanoparticle solutions can be estimated by measuring the contact angle between the rock surfaces and the oil droplet. The polished slice core samples were initially saturated with crude oil followed by saturation with a nanofluid solution. A drop of oil was then placed on the saturated slice core samples and the corresponding spreading of the droplet inside the rock surface was measured [28]. In this section, we will discuss two nanofluid systems, which include polymer–nanoparticles [28] and natural surfactant–polymer–nanoparticles solutions [3]. Figure 5.7 shows the variation in contact angle with time as the concentration of nanoparticles in the nanofluid solutions enhances from 0.1 to 0.5wt%. The wettability of

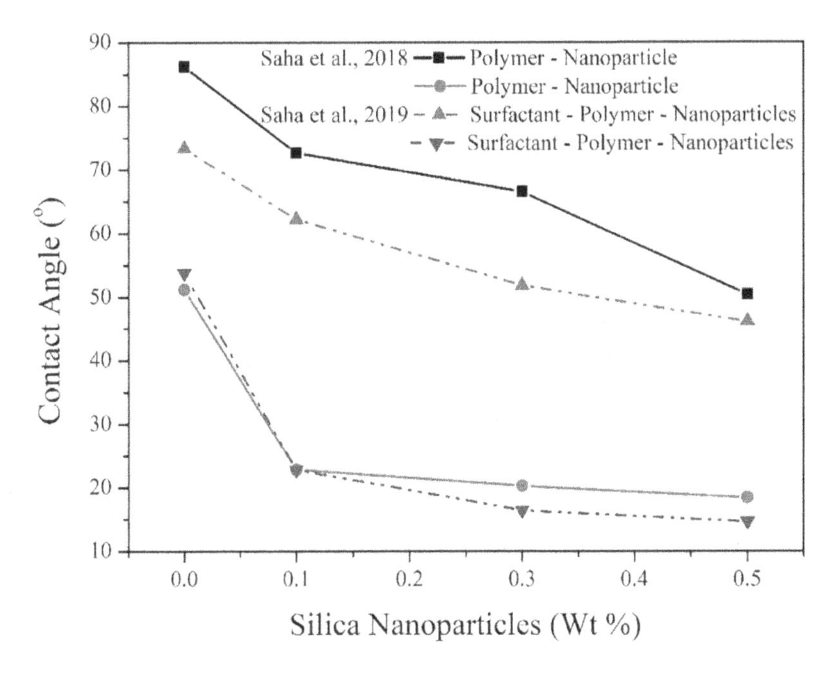

FIGURE 5.7
The changes in contact angle value with duration for nanofluid solutions at different concentrations of nanoparticles.

the reservoir in the absence of nanoparticles was found to be intermediate wet based on the change in the contact angle value from 86° to 51° for a polymer–nanoparticle and 73° to 53° for a surfactant–polymer–nanoparticle system. The introduction of nanoparticles changes the initial wettability of the reservoir from intermediate wet to water wet for both the nanofluid system. The contact angle reduces drastically with time as the concentration of nanoparticles in the nanofluid solutions increases. However, a small concentration of nanoparticles was sufficient enough to reduce the contact, which signifies the importance of nanoparticles towards wettability alteration. The lowest contact angle of 18° and 14° was observed for a polymer–nanoparticle and a surfactant–polymer–nanoparticle system, respectively, when the concentration of nanoparticles was at 0.5 wt%. The higher concentration of nanoparticles causes the drop to spread through the rock surface, which was induced by the large electrostatic repulsion between the nanoparticles [75]. Some authors also stated that alteration in wettability of the reservoir using nanoparticles could be due to the interaction between adsorbed nanoparticles at the interface and the carboxylate group present in the oil-saturated rock surface [76–78]. The nanoparticles which are found effective in changing the wettability of the rock surface are Al_2O_3, ZrO_2, SiO_2, TiO_2, MgO and Fe_2O_3 [45,79–81]. The wettability change depends on several factors like the types of nanoparticles, the concentration of nanoparticles, size of nanoparticles, temperature, salinity and properties of rock surfaces [82,83]. Hence, the mechanism of wettability alteration depends on the reservoir properties and careful investigation is necessary.

5.7 Nanofluid Flooding for Oil Recovery

After investigating all the mechanisms, the final flooding experiments were performed to determine the tertiary oil recovery by nanofluid EOR schemes. The flooding experiments were accompanied using Berea cores whose porosity and permeability varied from 25% to 27% and 700 to 1000 mD, respectively. The initial oil saturation in the rock was almost 80% and a slug size of 0.5 PV was selected based on the previously optimized parameters [84]. Figure 5.8a shows the cumulative recovery of crude oil using a chemical solution (the absence of nanoparticles) and nanofluid solutions. The recovery achieved by executing water flooding was in the range of 31%–33.5% and after inducing nanofluid (xanthan gum–silica nanoparticles) flooding, an additional oil recovery of 14.55%–21% was accomplished at a room temperature of 30°C. The enhancement in additional oil recovery continued till the concentration of nanoparticles reaches 0.3 wt% and under such a condition, a maximum cumulative oil recovery of around 55% was obtained. A further increase in the concentration of nanoparticles reduces the oil recovery, which is due to the blockage of porosity and permeability of the reservoir rock [29,35,85].

The highest oil recovery achieved was due to the combined effect of all the mechanisms studied above. Additionally, pressure drop encounter during the flooding experiments can further validate the oil recovery tendency (Figure 5.8b). A maximum pressure drop of 46.6 psi was reached when

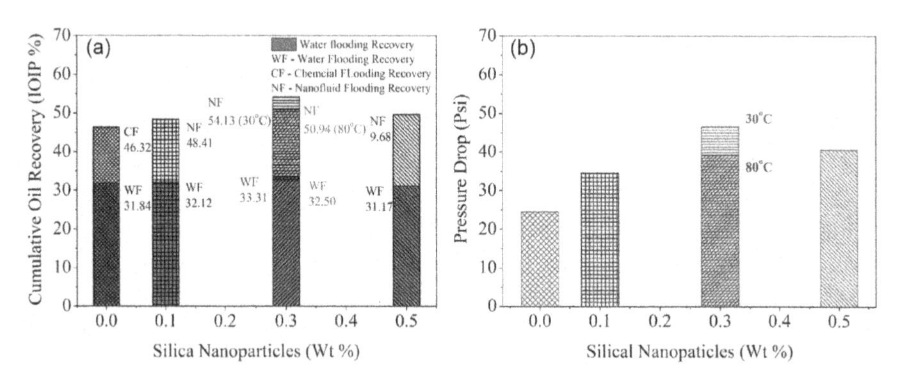

FIGURE 5.8
Chemical and nanofluid flooding data representing recovery of crude oil: (a) cumulative recovery of crude oil and (b) pressure drop encountered during the flooding experiments.

the concentration of nanoparticles was 0.3 wt%. The higher pressure drop indicates the establishment of a large oil bank that controls the recovery of crude oil [16,86,87]. Hence, an optimum nanoparticle concentration of 0.3 wt% was attained based on the flooding data. Moreover, for the same optimized system, the oil recovery reduced to around 19% as the temperature reached 80°C. This reduction in oil recovery due to temperature difference can be validated by the pressure drop data (Figure 5.8b). A lower pressure drop was detected, which specifies the disturbance in the formation of oil bank and hence lowers the recovery rate. Also, the lower oil recovery was expected because of the higher creaming rate of emulsion at an extreme temperature; however, the nanofluid/nanoparticles flooding were still successful to recover a good amount of residual crude oil.

5.8 Identification of Nanoparticles in Emulsion and Rock Surfaces

The adsorption of nanoparticles on the oil droplets and rock surfaces was identified by analyzing the emulsions and core samples before and after flooding experiments through FESEM characterization [28]. Figure 5.9 represents the core flooding setup and schematic images of the emulsion and Berea core after flooding experiments. The diagram represents adsorption of nanoparticles on the oil droplets and Berea rock surface in the form of clusters. These clusters were also encountered and validated with DLS characterization (Table 5.1).

The adsorption of these clusters on the rock surface has been investigated by several authors [3,29,88]. This adsorption has the potential to block the porosity and permeability, which causes the fluid to flow in the adjacent pores and thereby can enhance the sweep efficiency [9,89]. On the other hand, the negative impact of this behaviour can severely affect the flow of crude oil in the pores of the reservoir rock and therefore reduce the oil recovery [3,35,90]. Therefore, optimized conditions including the cost of nanoparticles should be formulated such that the recovery of oil can be maximized.

5.9 Nanofluid Field Projects and Technical Challenges

Nanofluid flooding has been attempted to a very limited scale and is currently in the development stage. So far, it has been implemented in some of the oil fields of China and Colombia. In China, Xinglongtai, Liaohe, Changqing Ansai, Jiangsu, Zhongyuan, Shengli, Wendong, Wangji and Henan oil fields

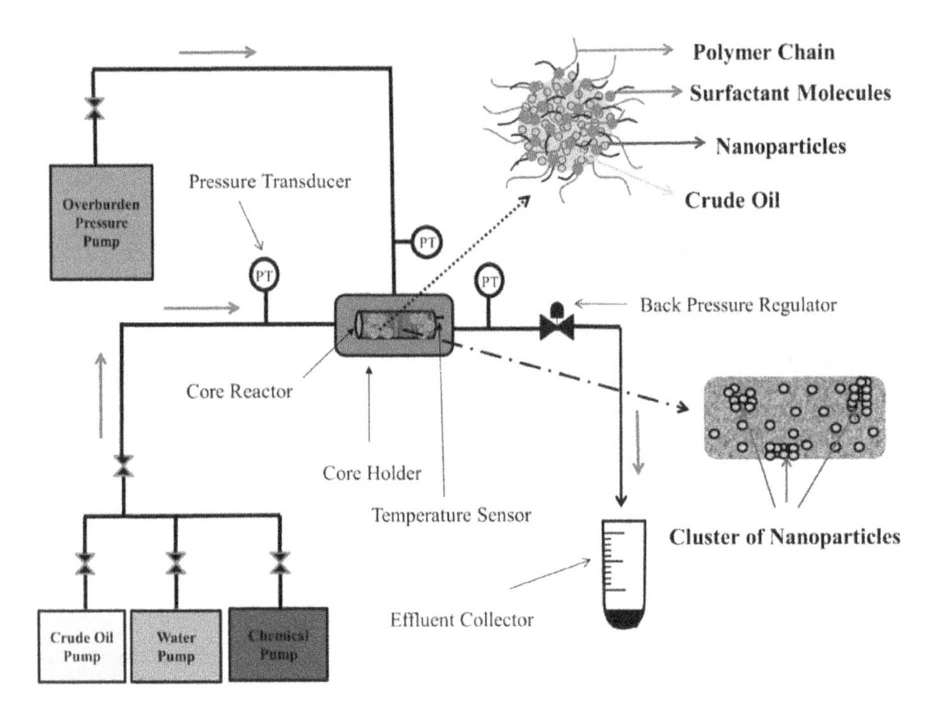

FIGURE 5.9
Schematic representation showing the presence of nanoparticles in oil droplets and cluster of nanoparticles in the Berea core along with a core flooding experimental setup.

are some of the places in which field test was executed [7,91–97]. The application of molecular deposition film as nanofluid flooding was impressive in the oil fields of Xinglongtai, Liaohe, Changqing Ansai and Jiangsu due to the wettability change mechanism and no swelling of minerals. The oil field of Xinglongtai and Liaohe was successful in enhancing the oil recovery factor by 1.68% which corresponds to 7092 t recovery [91,92]. The oil production in the Changqing Ansai field was similarly enhanced from an initial 0.78 t to a final production of 1.31 t [93]. The Jiangsu field likewise improved the displacement efficacy of oil to 27.9%, which was initially at 10% [94]. The implementation of SiO_2 nanoparticles in Zhongyuan, Wendong and Shengli oil fields alters the wettability of the rock due to which a decrease in the injection water pressure [91,95] and a higher pressure drop [7] were achieved. The polymer–SiO_2 microspheres were effective in enhancing the sweep efficiency in Changqing, Shengli, Wangji and Henan oil fields, which are found fruitful in recovering additional crude oil [91,95–97].

In Colombia, the oil fields of Chichimene and Castilla, which possess heavy crude oil, were investigated for nanofluid flooding [98]. The oil production in two different wells (CHSW26 and CH39) from the Chichimene field was enhanced by 310 and 87 bopd when 86 and 106 bbl of nanofluid solution were injected in those wells, respectively. Similar to the Castilla oil field, injecting 200 bbl nanofluid in CN154 well and 150 bbl nanofluid in CN147 well spikes the oil production by 270 and 280 bpd, respectively. The improved oil recovery was possible because of the viscosity reduction phenomena, which improved the mobility ratio. It was observed that after injecting the nanofluid, the mobility of oil continued to be the same for a period of 269 days. This showed the potential of nanofluid towards improved oil recovery in an economical way. The author also concluded that the cost of the project was recovered in a very short span of 4 months [98,99].

Several technical challenges have to be overcome for successful nanofluid flooding. Nanofluid flooding is costly in comparison to chemical flooding and it also includes the additional cost of emulsion separation as emulsions are stable. Therefore to reduce the cost of the operations, nanoparticles can be derived from cheap natural sources such as cellulose and starch. Currently, there exists limited research on the technical challenges of nanofluid EOR applications, which includes nanoparticles stability, separation of nanofluid from the produced water, reuse of the nanoparticles, cost efficiency, being an environmental pollutant, toxic, health risky and so on [88,99,100]. Nanofluid flooding showed promising results on a laboratory scale; however, the field applications are restricted due to these technical challenges. Additionally, the retention of nanoparticles on the rock surfaces for wettability change and improved sweep efficiency mechanisms depends on the rock morphology, which varies from reservoir to reservoir apart from the crude oil–nanoparticles chemistry. Hence, there exist huge scopes to explore and predict the success of nanofluid flooding in an oil field.

5.10 Conclusions

The potential of nanofluid flooding in the oil and gas industry to enhance residual oil recovery was discussed. In this work, a polymer–nanoparticle (nanofluid) system was chosen to describe the impact of nanofluid for improved oil recovery for heavy crude oil. The most important criteria of nanoparticles' stability in the nanofluid solution were analysed by visual sedimentation, particle size distribution and zeta potential values. The mechanisms which control the recovery factor such as IFT reduction, emulsion stability, rheology and wettability changes were investigated in detail. The coalescence rate of emulsion droplets was visualized by creaming index data and the formation of stable emulsions using nanoparticles was investigated by the size distribution of the droplets. Core flooding experiments were conducted and an additional oil recovery of 21% was obtained at a room temperature of 30°C. The potential of a nanofluid at a higher temperature of 80°C was confirmed by executing nanofluid flooding and an additional oil recovery of 19% was achieved. The data were further validated by the pressure drop data obtained during the flooding experiments. Moreover, field trial data of nanofluid flooding in addition to its technical challenges were addressed. Finally, the existing scope or area in which future research needs to be conducted for successfully implementing nanofluid flooding in the commercial platform was highlighted.

References

1. T. Singh, S. Shukla, P. Kumar, V. Wahla, V. K. Bajpai, and I. A. Rather, Application of nanotechnology in food science: Perception and overview, *Frontiers in Microbiology*, vol. 8, p. 1501, 2017. doi: 10.3389/fmicb.2017.01501.
2. O. Salata, Applications of nanoparticles in biology and medicine, *Journal of Nanobiotechnology*, vol. 2, 2004. doi: 10.1186/1477-3155-2-3.
3. R. Saha, R. Uppaluri, and P. Tiwari, Impact of natural surfactant (Reetha), polymer (Xanthan Gum), and silica nanoparticles to enhance heavy crude oil recovery, *Energy and Fuels*, vol. 33, pp. 4225–4236, 2019.
4. X. Y. Y, X. Y. Lin, and C. Y. Chen, Key factors influencing the toxicity of nanomaterials, *Chinese Science Bulletin*, vol. 58, pp. 2466–2478, 2013.
5. R. Dastjerdi and M. Montazer, A review on the application of inorganic nano-structured materials in the modification of textiles: Focus on anti-microbial properties, *Colloids Surfaces B*, vol. 79, pp. 5–18, 2010.
6. Y. Z. Ma, J. L. Fan, B. Y. Huang, X. Xiong, and D. L. Wang, Dispersion behavior of ultrafine/nanometer particulates in water medium, *Metallurgy & Metallurgical Engineering*, vol. 23, pp. 43–46, 2003.
7. Y. X. Wu, F. M. Ma, and M. F. Duan, Application efficiency and re-innovation counter measures of nanometer technology in oil and gas field development in China, *Drilling and Production Technology*, vol. 31, pp. 47–51, 2008.
8. J. T. Jiu, C. X. Li, C. F. Wang, B. H. Wang, and L. P. Li, Dispersion and application of nano-size ZnO, *Dyeing and Finishing*, vol. 1, pp. 1–13, 2002.
9. Y. Sun, D. Yang, L. Shi, H. Wu, Y. Cao, Y. He, et al., Properties of nano-fluids and their applications in enhanced oil recovery: A comprehensive review, *Energy and Fuels*, vol. 34, pp. 1202–1218, 2020.
10. H. Saboorian-Jooybari, M. Dejam, and Z. Chen, Heavy oil polymer flooding from laboratory core floods to pilot tests and field applications: Half-century studies, *Journal of Petroleum Science & Engineering*, vol. 142, pp. 85–100, 2016.
11. Z. Liu, Y. Li, J. Lv, B. Li, and Y. Chen, Optimization of polymer flooding design in conglomerate reservoirs, *Journal of Petroleum Science & Engineering*, vol. 152, pp. 267–274, 2017.
12. D. C. Standnes and I. Skjevrak, Literature review of implemented polymer field projects, *Journal of Petroleum Science & Engineering*, vol. 122, pp. 761–775, 2014.
13. M. Algharaib, A. Alajmi, and R. Gharbi, Improving polymer flood performance in high salinity reservoirs, *Journal of Petroleum Science & Engineering*, vol. 115, pp. 17–23, 2014.
14. R. L. Jewett and G. F. Schurz, Polymer flooding-a current appraisal, *Journal of Petroleum Technology*, vol. 22, pp. 675–684, 1970.

15. H. L. Chang, Polymer flooding technology yesterday, today, and tomorrow, *Journal of Petroleum Technology*, vol. 30, pp. 1–16, 1978.
16. H. Pei, G. Zhang, J. Ge, L. Zhang, and H. Wang, Effect of polymer on the interaction of alkali with heavy oil and its use in improving oil recovery, *Colloids and Surfaces A: Physicochemical and Engineering Aspects*, vol. 446, pp. 57–64, 2014.
17. W. Demin, Z. Zhang, L. Chun, J. Cheng, X. Du, and Q. Li, A pilot for polymer flooding of saertu formation S II 10–16 in the North of daqing oil field, in *SPE Asia Pacific Oil and Gas Conference*, 28–31 October, Adelaide, Australia, 1996, pp. 431–441.
18. E. Akiyama, A. Kashimoto, K. Fukuda, H. Hotta, T. Suzuki, and T. Kitsuki, Thickening properties and emulsification mechanisms of new derivatives of polysaccharides in aqueous solution, *Journal of Colloid and Interface Science*, vol. 282, pp. 448–457, 2005.
19. E. Akiyama, T. Yamamoto, Y. Yago, H. Hotta, T. Ihara, and T. Kitsuki, Thickening properties and emulsification mechanisms of new derivatives of polysaccharide in aqueous solution 2: The effect of the substitution ratio of hydrophobic/hydrophilic moieties, *Journal of Colloid and Interface Science*, vol. 311, pp. 438–446, 2007.
20. R. S. Seright, T. G. Fan, K. Wavrik, H. Wan, N. Gaillard, and C. Favero, Rheology of a new sulfonic associative polymer in porous media, *SPE Reservoir Evaluaton and Engineering*, vol. 14, pp. 726–734, 2011.
21. Y. Lu, W. Kang, J. Jiang, J. Chen, D. Xu, P. Zhang, et al., Study on the stabilization mechanism of crude oil emulsion with an amphiphilic polymer using the b-cyclodextrin inclusion method, *RSC Advances*, vol. 7, pp. 8156–8166, 2017.
22. W. M. Leung and D. E. Axelson, Thermal degradation of polyacrylamide and poly (acrylamide-coacrylate), *Journal of Polymer Science Part A*, vol. 25, pp. 1852–1864, 1987.
23. M. H. Yang, Rheological behavior of polyacrylamide solution, *Journal of Polymer Engineering*, vol. 19, pp. 371–381, 1999.
24. H. Kheradmand, J. Francois, and V. Plazanet, Hydrolysis of polyacrylamide and acrylic acid-acrylamide copolymers at neutral pH and high temperature, *Polymer*, vol. 29, pp. 860–870, 1988.
25. Q. Chen, Y. Wang, Z. Lu, and Y. Feng, Thermoviscosifying polymer used for enhanced oil recovery: Rheological behaviors and core flooding test, *Polymer Bulletin*, vol. 70, pp. 391–401, 2013.
26. G. Muller, Thermal stability of high molecular weight polyacrylamide aqueous solution, *Polymer Bulletin*, vol. 5, pp. 31–37, 1981.
27. B. F. Abu-Sharkh, G. O. Yahaya, S. A. Ali, E. Z. Hamad, and I. M. Abu-Reesh, Viscosity behavior and surface and interfacial activities of hydrophobically modified water-soluble acrylamide/N-phenyl acrylamide block copolymers, *Journal of Applied Polymer Science*, vol. 89, pp. 2290–2300, 2002.
28. R. Saha, R. Uppaluri, and P. Tiwari, Silica nanoparticle assisted polymer flooding of heavy crude oil: Emulsification, rheology, and wettability alteration characteristics, *Industrial and Engineering Chemistry Research*, vol. 57, pp. 6364–6376, 2018.
29. T. Sharma, S. Iglauer, and J. S. Sangwai, Silica nanofluids in an oilfield polymer polyacrylamide: Interfacial properties, wettability alteration, and applications for chemical enhanced oil recovery, *Industrial & Engineering Chemistry Research*, vol. 55, pp. 12387–12397, 2016.

30. T. Sharma, G. S. Kumar, B. H. Chon, and J. S. Sangwai, Thermal stability of oil-in-water Pickering emulsion in the presence of nanoparticle, surfactant, and polymer, *Journal of Industrial and Engineering Chemistry,* vol. 22, pp. 324–334, 2015.
31. B. P. Binks and A. Rocher, Effects of temperature on water-in-oil emulsions stabilised solely by wax microparticles, *Journal of Colloid and Interface Science,* vol. 335, pp. 94–104, 2009.
32. L. Hendraningrat, S. Li, and O. Torsaeter, A coreflood investigation of nanofluid enhanced oil recovery, *Journal of Petroleum Science and Engineering,* vol. 111, pp. 128–138, 2013.
33. H. Yousefvand and A. Jafari, Enhanced oil recovery using polymer/nanosilica, *Procedia Materials Science,* vol. 11, pp. 565–570, 2015.
34. A. Maghzi, R. Kharrat, A. Mohebbi, and M. H. Ghazanfari, The impact of silica nanoparticles on the performance of polymer solution in presence of salts in polymer flooding for heavy oil recovery, *Fuel,* vol. 123, pp. 123–132, 2014.
35. M. I. Youssif, R. M. El-Maghraby, S. M. Saleh, and A. Elgibaly, Silica nanofluid flooding for enhanced oil recovery in sandstone rocks, *Egyptian Journal of Petroleum,* vol. 27, pp. 105–110, 2018.
36. N. Kumar, T. Gaur, and A. Mandal, Characterization of SPN Pickering emulsions for application in enhanced oil recovery, *Journal of Industrial and Engineering Chemistry* vol. 54, pp. 304–315, 2017.
37. L. J. Duffus, J. E. Norton, P. Smith, I. T. Norton, and F. Spyropoulos, A comparative study on the capacity of a range of food-grade particles to form stable O/W and W/O Pickering emulsions, *Journal of Colloid and Interface Science,* vol. 473, pp. 9–21, 2016.
38. Z. Guo, M. Dong, Z. Chen, and J. Yao, A fast and effective method to evaluate the polymer flooding potential for heavy oil reservoirs in Western Canada, *Journal of Petroleum Science & Engineering,* vol. 112, pp. 335–340, 2013.
39. J. Wang and M. Dong, Optimum effective viscosity of polymer solution for improving heavy oil recovery, *Journal of Petroleum Science and Engineering,* vol. 67, pp. 155–158, 2009.
40. Y. Otsubo, Y. Cho, and C. Shi, Rheological properties of silica suspensions in polyacrylamide solutions, *Journal of Rheology,* vol. 28, pp. 95–108, 1984.
41. M. Kamibayashi, H. Ogura, and Y. Otsubo, Viscosity behavior of silica suspensions flocculated by associating polymers, *Journal of Colloid and Interface Science,* vol. 290, pp. 592–597, 2005.
42. N. K. Maurya and A. Mandal, Studies on behavior of suspension of silica nanoparticle in aqueous polyacrylamide solution for application in enhanced oil recovery, *Petroleum Science and Technology,* vol. 34, pp. 429–436, 2016.
43. A. E. Bayat, R. Junin, A. Samsuri, A. Piroozian, and M. Hokmabadi, Impact of metal oxide nanoparticles on enhanced oil recovery from limestone media at several temperatures, *Energy Fuels,* vol. 28, pp. 6255–6266, 2014.
44. M. Zallaghi, R. Kharrat, and A. Hashemi, Improving the microscopic sweep efficiency of water flooding using silica nanoparticles, *Journal of Petroleum Exploration and Production Technology,* vol. 8, pp. 1–11, 2017.
45. A. Karimi, Z. Fakhroueian, A. Bahramian, N. P. Khiabani, J. B. Darabad, R. Azin, et al., Wettability alteration in carbonates using zirconium oxide nanofluids: EOR implications, *Energy Fuels,* vol. 26, pp. 1028–1036, 2012.

46. A. Maghzi, A. Mohebbi, R. Kharrat, and M. H. Ghazanfari, Pore-scale monitoring of wettability alteration by silica nanoparticles during polymer flooding to heavy oil in a five-spot glass micromodel, *Transport in Porous Media*, vol. 87, pp. 653–664, 2011.
47. X. Y. Zhou, W. H. Li, and L. He, Dispersion stability of nanoparticles and its evaluation methods, *MatererProtect*, vol. 39, pp. 51–54, 2006.
48. X. L. Guo, S. G. Chen, and X. H. Pan, The effect of solvent on nano-SiC and Si_3N_4 dispersion, *Ceramic Engineering*, vol. 2, pp. 6–9, 2001.
49. S. J. Chung, J. P. Leonard, I. Nettleship, J. K. Lee, Y. Soong, D. V. Martello, et al., Characterization of ZnO nanoparticle suspension in water: Effectiveness of ultrasonic dispersion, *Powder Technology*, vol. 194, pp. 75–80, 2009.
50. J. T. Jiu, C. X. Li, C. F. Wang, B. H. Wang, and L. P. Li, Dispersion and application of nano-size ZnO, *Dyeing Finising*, vol. 1, pp. 1–3, 2002.
51. X. L. Song, Z. H. Yang, G. Z. Qiu, and X. H. Qu, Synthesis of Al_2O_3 nanoparticles and dispersion and stability for its suspension, *Journal of Central South University of Technology*, vol. 34, pp. 484–488, 2003.
52. J. Wang and Y. F. Xu, On the dispersion and stability of ultrafine powders in liquid, *Journal of Hefei University of Technology*, vol. 25, pp. 123–126, 2002.
53. U. Farooq, M. T. Tweheyo, J. Sjöblom, and G. Øye, Surface characterization of model, outcrop, and reservoir samples in low salinity aqueous solutions, *Journal of Dispersion Science and Technology*, vol. 32, pp. 519–531, 2011.
54. J. Ren, S. Song, A. Lopez-Valdivieso, J. Shen, and S. Lu, Dispersion of silica fines in water-ethanol suspensions, *Journal of Colloid and Interface Science*, vol. 238, pp. 279–284, 2001.
55. Y. H. Wei, X. T. Wang, H. L. Liu, and Y. Hu, Rheological behaviour of superfine γ-Al_2O_3 suspension, *Journal of East China University of Science and Technology*, vol. 25, pp. 518–520, 1999.
56. D. Harbottle, P. Bueno, R. Isaksson, and I. Kretzschmar, Coalescence of particle-laden drops with a planar oil–water interface, *Journal of Colloid and Interface Science*, vol. 362, pp. 235–241, 2011.
57. M. Wisniewska, The temperature effect on the adsorption mechanism of polyacrylamide on the silica surface and its stability, *Applied Surface Science*, vol. 258, pp. 3094–3101, 2012.
58. M. Tarek, Investigating nano-fluid mixture effects to enhance oil recovery, SPE-178739-STU, *in SPE Annual Technical Conference and Exhibition, Society of Petroleum Engineers: Richardson*, September 28–30, Houston, Texas, 2015.
59. R. Hashemi, N. N. Nassar, and P. Pereira Almao, Enhanced heavy oil recovery by in situ prepared ultradispersed multimetallic nanoparticles: A study of hot fluid flooding for athabasca bitumen recovery, *Energy and Fuels*, vol. 27, pp. 2194–2201, 2013.
60. E. A. Taborda, C. A. Franco, S. H. Lopera, V. Alvarado, and F. B. Cortés, Effect of nanoparticles/nanofluids on the rheology of heavy crude oil and its mobility on porous media at reservoir conditions, *Fuel*, vol. 184, pp. 222–232, 2016.
61. H. Patel, S. Shah, and R. Ahmed, Effects of nanoparticles and temperature on heavy oil viscosity, *Journal of Petroleum Science and Engineering*, vol. 167, pp. 819–828, 2018.
62. K. Almdal, J. Dyre, S. Hvidt, and O. Kramer, Towards a phenomenological definition of the term 'gel', *Polymer Gels and Networks*, vol. 1, pp. 5–17, 1993.

63. W. Kang, B. Xu, Y. Wang, Y. Li, X. Shana, F. An, et al., Stability mechanism of W/O crude oil emulsion stabilized by polymer and surfactant, *Colloids and Surfaces A: Physicochemical Engineering Aspects*, vol. 384, pp. 555–560, 2011.
64. Y. Kazemzadeh, E. S. Eshraghi, K. Kazemi, S. Sourani, M. Mehrabi, and Y. Ahmadi, Behavior of asphaltene adsorption onto the metal oxide nanoparticle surface and its effect on heavy oil recovery, *Industrial and Engineering Chemistry Research*, vol. 54, pp. 233–239, 2015.
65. A. Roustaei, S. Saffarzadeh, and M. Mohammadi, An evaluation of modified silica nanoparticles' efficiency in enhancing oil recovery of light and intermediate oil reservoirs, *Egyptian Journal of Petroleum*, vol. 22, pp. 427–433, 2013.
66. E. C. Donaldson, G. V. Chilingarian, and T. F. Yen, Eds., *Enhanced Oil Recovery II, Processes and Operations*. Elsevier Science Publ. Co., New York, 1989.
67. A. Samanta, K. Ojha, A. Sarkar, and A. Mandal, Surfactant and surfactant-polymer flooding for enhanced oil recovery, *Advances in Petroleum Exploration and Developmen*, vol. 2, pp. 13–18, 2011.
68. Y. H. Cao, E. Dickinson, and D. J. Wedlock, Creaming and flocculation in emulsions containing polysaccharide, *Food Hydrocolloids*, vol. 4, pp. 185–195, 1990.
69. E. Dickinson, J. G. Ma, and M. J. W. Povey, Creaming of concentrated oil in-water emulsions containing xanthan, *Food Hydrocolloids*, vol. 8, pp. 481–497, 1994.
70. A. K. Ghosh and P. Bandyopadhyay, Polysaccharide-protein interactions and their relevance in food colloids, in Karunaratne, D. N. (Ed.), *The Complex World of Polysaccharide*. Intech Open Science Open Minds, Rijeka, Croatia, pp. 395–408, 2012.
71. B. P. Binks and C. P. Whitby, Silica particle-stabilized emulsions of silicone oil and water: Aspects of emulsification, *Langmuir*, vol. 20, pp. 1130–1137, 2004.
72. L. W. Lake. *Enhanced Oil Recovery*. SPE, Houston, TX, 2010.
73. S. Iglauer, M. Sarmadivaleh, C. Geng, and M. Lebedev, In-situ residual oil saturation and cluster size distribution in sandstones after surfactant and polymer flooding imaged with X-ray micro-computed tomography, *Presented at the International Petroleum Technology Conference*, January 19–22, Doha, Qatar, 2014.
74. M. Arhuoma, M. Dong, D. Yang, and R. Idem, Determination of water-in-oil emulsion viscosity in porous media, *Industrial & Engineering Chemistry Research*, vol. 48, pp. 7092–7102, 2009.
75. S. Li, L. Hendraningrat, and O. Torsaeter, Improved oil recovery by hydrophilic silica nanoparticles suspension: 2 phase flow experimental studies, *Presented at the International Petroleum Technology Conference, IPTC-16707*, Beijing, China, 2013.
76. A. D. Monfared, M. H. Ghazanfari, M. Jamialahmadi, and A. Helalizadeh, Adsorption of silica nanoparticles onto calcite: Equilibrium, kinetic, thermodynamic and DLVO analysis, *Chemical Engineering Journal*, vol. 281, pp. 334–344, 2015.
77. A. D. Monfared, M. H. Ghazanfari, M. Jamialahmadi, and A. Helalizadeh, Potential application of silica nanoparticles for wettability alteration of oil–wet calcite: A mechanistic study, *Energy Fuels*, vol. 30, pp. 3947–3961, 2016.
78. A. Maghzi, S. Mohammadi, M. H. Ghazanfari, R. Kharrat, and M. Masihi, Monitoring wettability alteration by silica nanoparticles during water flooding to heavy oils in five-spot systems: A pore-level investigation, *Experimental Thermal and Fluid Science*, vol. 40, pp. 168–176, 2012.

79. L. Hendraningrat and O. Torsæter, Metal oxide-based nanoparticles: Revealing their potential to enhance oil recovery in different wettability systems, *Applied Nanoscience*, vol. 5, pp. 181–199, 2014.
80. R. N. Moghaddam, A. Bahramian, Z. Fakhroueian, A. Karimi, and S. Arya, Comparative study of using nanoparticles for enhanced oil recovery: Wettability alteration of carbonate rocks, *Energy Fuels*, vol. 29, pp. 2111–2119, 2015.
81. E. Joonaki and S. Ghanaatian, The application of nanofluids for enhanced oil recovery: Effects on interfacial tension and coreflooding process, *Petroleum Science and Technology*, vol. 32, pp. 2599–2607, 2014.
82. R. Aveyard, B. P. Binks, and J. H. Clint, Emulsions stabilized solely by colloidal particles, *Advances in Colloid and Interface Science*, vol. 100, pp. 503–546, 2003.
83. J. A. Ali, K. Kolo, A. K. Manshada, and A. H. Mohammadid, Recent advances in application of nanotechnology in chemical enhanced oil recovery: Effects of nanoparticles on wettability alteration, interfacial tension reduction, and flooding, *Egyptian Journal of Petroleum*, vol. 27, pp. 1371–1383, 2018.
84. R. Saha, R. V. S. Uppaluri, and P. Tiwari, Influence of emulsification, interfacial tension, wettability alteration and saponification on residual oil recovery by alkali flooding, *Journal of Industrial and Engineering Chemistry*, vol. 59, pp. 286–296, 2018.
85. H. Ehtesabi, M. M. Ahadian, V. Taghikhani, and M. H. Ghazanfari, Enhanced heavy oil recovery in sandstone cores using TiO_2 nanofluids, *Energy and Fuels*, vol. 28, pp. 423–430, 2014.
86. L. Chen, G. Zhang, J. Ge, P. Jiang, J. Tang, and Y. Liu, Research of the heavy oil displacement mechanism by using alkaline/surfactant flooding system, *Colloids and Surfaces A: Physicochemical and Engineering Aspects*, vol. 434, pp. 63–71, 2013.
87. H. Pei, G. Zhang, J. Ge, L. Jin, and C. Ma, Potential of alkaline flooding to enhance heavy oil recovery through water-in-oil emulsification, *Fuel*, vol. 104, pp. 284–293, 2013.
88. A. O. Gbadamosi, R. Junin, M. A. Manan, A. Agi, and A. S. Yusuff, An overview of chemical enhanced oil recovery: Recent advances and prospects, *International Nano Letters*, vol. 9, pp. 171–202, 2019.
89. J. Gong, W. B. Xu, and H. H. Tao, Research status of nano liquid flooding technology, *Natural Gas Industry*, vol. 26, pp. 105–107, 2006.
90. N. A. Ogolo, O. A. Olafuyi, and M. O. Onyekonwu, Enhanced oil recovery using nanoparticles, SPE-160847-MS, in *SPE Saudi Arabia Section Technical Symposium and Exhibition*, 8–11 April, Al-Khobar, Saudi Arabia, 2012.
91. Y. C. Ke and G. Y. Wei, Application and development of nanomaterials in oil drilling and recovery, *Oilfield Chemistry*, vol. 25, pp. 189–192, 2008.
92. X. Zhao, Status quo of dispersed flooding with magnetic nanoparticle inclusion, *Drilling and Production Technology*, vol. 40, pp. 95–98, 2017.
93. J. L. Teng, C. H. Wang, Y. M. Han, Z. W. Liu, and Y. F. Yang, Research and application of nanometer diaphragm flooding technology in An Sai Wang 20–8 oil reservoir, *Petroleum Geosciences and Engineering*, vol. 22, pp. 99–102, 2008.
94. W. H. Yao and L. Yu, Experiment research on flooding by nano-solution in Jiangsu oil field, *Petrochemical Applications in Industries*, vol. 30, pp. 8–12, 2011.
95. Z. P. Cheng, Research on the technology of nano: Assisted oil displacement in the two - element flooding system of Da Gang Guan 109-1 oil reservoir, Master's Thesis, China University of Petroleum (Beijing), Beijing, China, 2016.

96. Y. F. Cai, X. R. Li, M. Q. Shi, L. H. Yang, X. C. Liu, and F. P. Wu, Research on the adaptability of polymeric nanospheres flooding in extra-low permeability reservoir in changqing oilfeld, *Oil Drilling & Production Technology*, vol. 35, pp. 88–93, 2013.
97. Y. Chen, Y. Q. Sun, D. L. Wen, X. Tian, and S. L. Gao, Evaluation and application on profile control of polymer nano-microspheres, *Petroleum Drilling Technology*, vol. 40, pp. 102–106, 2012.
98. R. Zabala, C. Franco, and C. Fortes, Application of nanofluid for improving oil mobility in heavy oil and extra-heavy oil: A field test, SPE-179677-MS, *Presented at the SPE Improved Oil Recovery Conference*, 11–13 April, Tulsa, Oklahoma, 2016.
99. A. Agi, A. Junin, and A. Gbadamosi, Mechanism governing nanoparticle flow behaviour in porous media: Insight for enhanced oil recovery applications, *International Nano Letters*, vol. 8, pp. 49–77, 2018.
100. C. Negin, S. Ali, and Q. Xie, Application of nanotechnology for enhancing oil recovery: A review, *Petroleum*, vol. 2, pp. 324–333, 2016.

6

Problems and Challenges in Chemical EOR

6.1 Introduction

The application of chemical EOR for higher residual oil recovery is of enormous importance in order to overcome the energy demand. The important characteristics of successful chemical EOR such as crude oil properties, reservoir heterogeneity, salinity, temperature, chemical formulations, adsorption, different mechanisms and the economy of the process design have already been discussed in the earlier chapters of the book. The restrictions encountered during chemical flooding schemes (alkali, surfactant, polymer, nanoparticles and their combinations) are discussed in this chapter. The technical difficulties such as precipitation, scaling, corrosion, low chemical injectivity, plugging of wells, damage of wellbore system, chemical degradation, polymer dissolution, separation of produced fluids, bacterial growth, logistics and handling issues can be encountered during the implementation of chemical EOR in the fields [1–4]. Therefore, all these limitations are identified and discussed thoroughly to understand the limitation of the system. After discussing the limitation, technical solutions will be discussed which will highlight the important factors to be considered to achieve successful pilot or field chemical EOR projects. Additionally, a summary of some of the latest laboratory and field chemical EOR projects in different countries are described.

6.2 Limitations of Chemical EOR

The limitations or problems faced during chemical EOR including precipitation, scaling, formation damage, difficulties associated with the produced emulsion, chromatographic separation, issues of logistic and handling, cost and others are described in this section. These are the important factors that may affect the economy of the process and at the same time can lead to an industrial hazard [1,4–6].

6.2.1 Precipitation and Scaling

Alkali, when injected into the reservoir, reacts with the rock surface dissolving some of the rock components. These dissolved rock components are then carried to the production well by the flooding liquid. As a consequence of this produced liquid and decrease in the temperature and pressure, precipitation and deposition occur in the pipeline, tubing, pumps, etc. Alkali additionally reacts with the divalent ions present in the reservoir brine resulting in further precipitation. This results in scaling and plugging of the wellbore, which leads to the breakdown of the process.

The salinity of formation water varies from reservoir to reservoir with the majority of cations and anions being Na^+, K^+, Ca^{2+}, Mg^{2+} and Cl^-, HCO_3^-, CO_3^{2-}, SO_4^{2-}, respectively. SO_3^{2-} ions are formed by the interaction of alkali with the formation minerals. The interaction between injected alkali, ions and rock minerals results in the formation of hydroxyl, carbonate, sulfate and silicate scales [1,5].

$$\text{Hydroxyl scales}: Ca^{2+} + 2OH^- = Ca(OH)_2 \downarrow;\ Mg^{2+} + 2OH^- = Mg(OH)_2 \downarrow \tag{6.1}$$

$$\text{Carbonate scales}: Ca^{2+} + CO_3^{2-} = CaCO_3 \downarrow;\ Mg^{2+} + CO_3^{2-} = MgCO_3 \downarrow \tag{6.2}$$

$$\text{Sulfate scales}: Ca^{2+} + SO_4^{2-} = CaSO_4 \downarrow;\ Mg^{2+} + SO_4^{2-} = MgSO_4 \downarrow \tag{6.3}$$

$$\text{Silicate scales}: Ca^{2+} + SiO_3^{2-} = CaSiO_3 \downarrow;\ Mg^{2+} + SiO_3^{2-} = MgSiO_3 \downarrow \tag{6.4}$$

Another scale formation observed due to divalent ions present in hard brine is the surfactant precipitation that results in surfactant retention [1]. The general reaction involved is as follows:

$$2P^- + Q^{2+} \rightarrow QP_2 \downarrow \tag{6.5}$$

where P^- indicates anionic surfactant and QP_2 is the divalent ions and surfactant complex with negligible solubility in brine.

A monovalent cation is formed when a multivalent cation reacts with an anionic surfactant at lower hardness. This monovalent then exchanges the cationic site present in the rock surface with the reaction as given below [1,7].

$$P^- + Q^{2+} \rightarrow QP^+ \tag{6.6}$$

$$N - Clay + QP^+ \rightarrow QP - Clay + Na^+ \tag{6.7}$$

6.2.2 Formation Damage

The permeability of the reservoir formation, which is considered as the ability of the fluid to flow through porous media, is affected because of the retention of chemicals on the rock surface or due to erosion of the formation rock by alkali interaction. Thus, the overall combined effect leads to the damage of the formation rock. The polymer retention mechanisms such as hydrodynamic retention, mechanical entrapment and adsorption behaviour causing inaccessible pore volume occur due to polymer flocculation at high salinity which plugs the rock pores [8,9]. This behaviour in the small pores of the reservoir rock is known as the particular filtration phenomenon. Similarly, during alkali flooding, the dissolved rock and minerals together migrate in the pores of the rock and block the pore throats thus reducing the permeability. The lower permeability is severely affected as the pore size in such case is small and can easily be blocked [5]. This damage of formation therefore can severely reduce the oil recovery factor and additionally increase the technical and economic challenges during oil field execution [9,10].

Similarly, in nanofluid core flooding experiments, nanoparticles are adsorbed on the rock surface, which blocks the pores of the reservoir rock thus ultimately reducing the permeability [11–15]. Though chemical nanofluid in general increases the oil recovery, enhancing the nanoparticles concentration beyond optimum can significantly reduce the permeability which ultimately reduces the tertiary oil recovery factor [11,12]. Therefore, this phenomenon of pore blockage by nanoparticles has to be incorporated in addition to various problems and challenges such that successful chemical nanofluid EOR can be formulated in the future for oil field operations.

6.2.3 Produced Emulsion Treatment

Chemicals when injected into oil reservoirs react with the crude oil to form stable emulsions. These stable emulsions can also be formed during water injection because of asphaltene, a natural emulsifier component present in crude oil [3]. Emulsion increases the sweep efficiency to enhance residual oil recovery but it may raise problems with transportation and separation of oil and water at the production platform. The existence of stable emulsions in the produced fluid creates difficulties in separation and processing them in the separator [16]. The addition of a surfactant and polymer along with in-situ soap formed by alkali–oil chemistry enhances emulsion stability to a greater extent due to electrostatic and steric effect phenomena [16,17], which are difficult to demulsify. In 1992, during an ASP pilot plant operation at Shengli Oil Field China, it was difficult to separate oil and water even after using weak alkali, Na_2CO_3. Furthermore, though it was not possible to treat the produced water, they had to re-inject the produced water back into the reservoir [5]. In the Chinese oil field, tremendous scaling and emulsion difficulties were detected due to the use of strong alkali NaOH. Hence, to overcome such difficulties, an alkali-free chemical EOR was planned.

A demulsifier is generally used to separate the stable emulsion by getting adsorbed and spreading at the oil–water interface to form a film thus affecting the interfacial intensity. Demulsifiers such as SP169 (linear molecule), AE1910 (multi-branch chain molecules) and JS-8 (complex type) were examined [3,18]. Linear molecules are accountable to get adsorbed at the interface and spread over to form a film, whereas branch molecules affect the interfacial intensity. JS-8 which is of complex type behaves either as linear or branched molecules [3,19]. There are various other demulsifiers such as FD408-01 and GFD310-10 whose selection depends on the type of chemicals used for EOR processes [20].

6.2.4 Chemical Separation

The adsorption of chemicals on the reservoir rock is one of the important parameters to be considered during chemical flooding. Therefore in order to understand the adsorption characteristics of alkali, surfactant or polymer during complex ASP chemical flooding, the reservoir should be considered as a single large column and then measure the breakthrough of chemicals to identify the adsorption behaviour. Adsorption of these chemicals can severely impact the synergy effect which can increase the IFT affecting the oil displacement. Thus, it is important to remember that the chromatographic separation of chemical should be minimized during the displacement process so that oil recovery can be enhanced.

A study on chromatographic separation of chemicals during alkali–surfactant–polymer (ASP) flooding was investigated in Daqing Oil Field, China. In the core flood test, they observed the breakthrough of chemicals in sequence as polymer, alkali and surfactant. The breakthrough time for polymer was fast as the polymer cannot access the small pore volume. The retention of surfactant was highest, which is directly proportional to the clay content of the reservoir rock. They used bio-surfactant, which reduces the adsorption amount of expensive surfactant and hence the chemical cost, was lowered [3,5,21].

6.2.5 Water Disposal Treatment and Facility Problems

The produced water from the field has to undergo treatment before its disposal to meet the disposal standards. The existence of stable emulsion in the produced water is one of the major concerns as it contains oily and suspended solids. The adsorption of injected chemicals on the oil surface and the suspended solids in the produced water restrict it from the disposal. Therefore the overall process to meet the standard for disposal of the produced water becomes strenuous [1,9,16].

The facility problems were observed while handling fluid for chemical EOR processes [3]. Problems such as pump depreciation and reduction in working life span are reported. The scaling which occurs during chemical (ASP) flooding can severely decrease the life span of the screw pump as observed in Daqing oil field, China. The average pump life was reduced to 97 days, which was 375 and 618 days for polymer and water flooding respectively [20]. Additional problems observed with the facility while handling polymer solutions are because of its viscoelastic behaviour. When the polymer solution flows from the main supply to branch routes, a pulling force tries to pull the polymer back in the mainline and this happens because of the viscoelastic behaviour of the polymer. As the velocities in the main line and branch route increase, the magnitude of this pulling force enhances. This results in oscillation which changes the extension viscosity and normal stress thus causing the pump to vibrate. Therefore the solution to this problem is to increase the size of the pipe [22]. A large dead area is created at the bottom of the maturation tank while mixing polymer solution, which makes the mixing process difficult and further consumes more energy. To solve this issue centralizers are used [23].

6.2.6 Challenges in Offshore Oil Field

The challenges in offshore fields during chemical EOR applications involves remote location, limited workspace in the deck and large well spacing [24]. Table 6.1 summarizes all these difficulties and their effects on chemical EOR processes in detail [1,24]. If the offshore production well is located in a remote area, it becomes difficult to shift and storage the chemicals, thus making the process expensive [25]. The lesser space in the deck due to the involvement of floating, production, storage and offloading (FPSO) vessel is another issue faced during oil production from an offshore field. The large well spacing, which arises due to high good drilling, especially affects the monitoring and controlling of the ASP chemical EOR process in the offshore platform. Also, since the incremental residual oil is observed after certain years of chemical injection, the pay-out period for offshore reservoir is long.

6.2.7 Cost of Chemicals

Chemical EOR performed on a laboratory scale requires a small amount of chemicals and hence does not indicate any cost-effectiveness of the system. The price of chemicals is a major factor that decides the process economy as the amount of chemicals as well as cost increases enormously for field applications. In general, the execution of any EOR schemes is decided based on the process efficiency to recover residual oil from reservoirs in an economical way. However, the involvement of chemical EOR is decided based on the market oil price, which is decided by the OPEC countries. Chemical EOR is advisable for the field implementation during higher oil price in the market such that the oil industries can sustain their profits. A lot of chemical EOR projects in the year 2015 had to stop to prevent losses as the oil price dropped down to $ 40/bbl. In an oil field in China (Daqing), to maintain the

TABLE 6.1

The Challenges and Process Effects Observed during Offshore Chemical EOR [1,24]

	Process Effects				
Challenges	**Subsurface Efficiency**	**Logistics**	**Injection**	**Production**	**Environment**
Remote locations		Surfactant availability and cost		On-site separation preferred	
Poor weather	Effect of interrupted supply	Large storage requirements	Reliability, intermittent injection		Accidental chemical releases
Expensive wells	Fewer injection wells to monitor the process	Distribution of ASP solutions over long distances	Produced water reinjection likely		No disposal well for produced water
Space weight	Low adsorbing surfactants	High-active chemicals		Space-efficient separation equipment	
Seawater disposal	Salt tolerant surfactant Effect of reinjection on chemical efficacy	Recycle surfactants	Water softening and deaeration Produced water reinjection		Overboarding toxic surfactants

oil production economically by a chemical (ASP) EOR process, weak alkali is substituted by strong alkali to reduce the overall cost of the operation [9,26].

Furthermore, the involvement of nanoparticles, which are proven to be effective for higher oil recovery at a laboratory scale, may not be effective for commercialization purpose. This could be due to the requirement of a higher quantity of nanoparticles for field applications and the cost associated with it. Therefore, nanoparticles possessing higher efficiency can be derived from cheap sources to make the process economically viable for field applications. Additionally, enormous experimental data considering real-field scenario have to be generated to establish an accurate model or correlation for their applicability on a large scale.

6.3 Case Studies on Challenges of Chemical EOR

One of the important case studies on the challenges (scaling) faced during the ASP flooding field test and the actions taken to overcome is discussed. The injection of ASP slug in Daqing oil reservoirs (China) did face severe issues like scaling, corrosion, injectivity, chemical degradation, pump failure, reducing the lifespan of down-hole and lifting instruments, emulsions in produced fluids, reduced efficiency and additional operational cost [1,3,27,28]. The scaling was detected majorly due to the presence of silicon dioxide and carbonate in the produced water [28]. Studies performed earlier on Daqing fields stated that strong alkalis such as NaOH could result in greater scaling as compared to weak alkali Na_2CO_3 [29–31]. The strong alkali in ASP was superior as it resulted in ultra-low IFT and higher oil recovery with respect to weak alkali; however, weak alkali ASP was preferred in large-scale projects considering the process economics. Therefore in the field test, strong alkali in ASP flooding was replaced by weak alkali which showed lower scaling, improved injectivity, better productivity and lower cost. Out of the 30 field tests conducted in the China oil field (except Daqing), all the fields were exposed to either Na_2CO_3 or NaOH–Na_2CO_3 mixture-based ASP flooding to overcome scaling issues [29]. They also reported that increased oil recovery is not the best parameter for successful ASP flooding, it is the oil price (economics) that should be considered.

6.4 Technical Solutions for Chemical EOR

The technical solutions to perform effective chemical EOR schemes are described in this section. The important factors to be considered involves adsorption inhibitors, scale inhibitors, precipitation inhibitors, water treatment in chemical EOR, transport, storage, cost and environmental concern.

Chemical Adsorption Inhibitors: The retention of surfactant by adsorption in the reservoir rock can be minimized by the applications of adsorption inhibitors. The inhibitors also known as sacrificial agents are highly recommended during chemical EOR processes [32]. The sacrificial agent reacts with the monovalent, divalent and polyvalent cations that exist in the brine forming complexes, thus reducing the available cations for chemicals to interact. Additionally, due to high surface coverage and the low desorption rate of the sacrificial agent, the active adsorption sites on the reservoir rocks were reduced. This process assists in maintaining a proper chemical slug reducing the effect of chemical dilution. Some of the sacrificial agents that are scrutinized and found effective are lignosulfonates and polyacrylate [9,32–34]. The cost of the sacrificial agent is one of the important factors to be considered to make the process economically feasible.

Scale Inhibitors: The chemicals during flooding experiments react with the divalent cations present in the formation water forming scales. This scale blocks the flow of fluid inside the pipelines and reduces the cumulative oil recovery. The porosity and permeability of the reservoir rock are also affected during such processes [9]. Therefore, scale inhibitor chemicals are employed which prevent scaling when added in small quantity in the scaling water. The role of scale inhibitors is to get adsorbed on the crystal surface, which restricts the growth of tiny crystals thus avoiding the formation of scales in metals. It also avoids the crystal from adhering into the pipelines due to the coating of the formed crystal [35,36]. The ratio of surface area to volume/mass ratio is the factor that decides the efficiency and speed at which the chemical inhibitor removes the scale [37].

The conventional scale inhibitors are soluble in water because they are hydrophilic in nature. Organic polymers, organophosphorus compounds and inorganic phosphates are some of the common conventional scale inhibitors. In the oil and gas industry, the most commonly used commercial-scale inhibitors are diethylenetriaminepenta methylene phosphonic acid (DETPMP) and polyphosphono carboxylic acid (PPCA). DETPMP works on the inhibition mechanism of crystal growth inhibition and PPCA by nucleation [1,38,39]. Green-scale inhibitors are those chemicals that are used to reduce environmental issues, their performance is economical with social responsibilities [40,41]. The examples of green-scale inhibitors are maleic acid, inulin, poly-alpha, beta-D, L-aspartate and pteroyl-L-glutamic acid [1]. The implementation of these types of green-scale inhibitors in the oil and gas industry has not been explored [42], though some alternatives have been proposed [43].

Precipitation Inhibitors: Sodium acrylate are formed in-situ when alkali and acrylic acid are introduced in the injection water [44]. The amount of sodium acrylate formed depends on the reaction between the acrylic acid and the sodium ions of the alkali (sodium carbonate). The adsorption of sodium acrylate on the growing active site of the metal cations avoids the precipitation process. The reaction of the precipitation process is [1,44]

$$C_3H_4O_2 + Na \rightarrow C_3H_3NaO_2 + H_2O \tag{6.8}$$

$$C_3H_4O_2 + Na_2CO_3 \rightarrow C_3H_3NaO + CO_2 + H_2O \tag{6.9}$$

The ionic strength of the formation water or brine solution changes with sodium acrylate, which accelerates the solubility of magnesium and calcium ions [45]. Therefore, the surfactant when injected in the reservoirs does not show any precipitation activity when the calcium and magnesium ions are eliminated from the aqueous phase [46]. Also because of these in-situ inhibitors, the hard formation water can be directly used without softening, thus reducing the cost and environmental hazards of chemical EOR processes [1].

Pre-Flush of Reservoir: The reservoirs possessing hard brine are not favourable for chemical EOR processes as in such cases, the performance of chemicals is affected due to water hardness. Therefore to defeat this challenge, researchers have overcome with a solution of pre-flushing the reservoir before injecting chemicals slug into the reservoirs. The pre-flush is usually considered as conditioning the reservoirs towards the desirable condition. The pre-flush is a formulated water slug with exceptional characteristics that reduce the hardness of brine prior to optimum formulated chemical slug injection. The parameters to be considered for chemical EOR schemes includes total dissolved solids of brine, pre-flush salinity, chemical concentration, slug size of pre-flush and chemicals [47,48].

Water Treatment of Chemicals: The design of optimum alkali–surfactant–polymer flooding for any reservoirs should consider the quantity of divalent ions present in the formation water. This divalent ion in the brine reacts with the chemicals to form carbonate and hydroxide undergoing precipitation and ultimately enhances the scaling. Therefore, the formation has to be treated to reduce the concentration of Ca^{2+} and Mg^{2+} ions. This process of softening the brine can be overcome by treating the hard water with strong or weak acid cation resins (SAC or WAC). The above method is applicable only when the concentration of oil or grease in the feed water is <50 ppm. If oil or grease content is high (>500 ppm) then the select ion sequestration (SIS) method can be employed [1].

Logistic Issues: The logistics (storage and transportation cost) department especially exploring in offshore oil fields plays an important role while evaluating the economic feasibility of the process. The storage requirements of aqueous chemicals solution during EOR applications contribute towards the additional cost. This can be overcome by considering surfactants that have low absorptivity and higher activity such that the additional storage requirements can be minimized. Furthermore, the formulation of a chemical slug that avoids the use of co-solvent can also reduce the logistic chemical charges [49]. The effort in reducing the logistic cost by changing the chemistry of the chemicals should be overcome by other changes such that efficient oil recovery can be obtained.

Derived Cost-Effective Chemicals for EOR Applications: The cost of the chemical has a huge impact on the process and hence researchers have focussed on deriving chemicals from a cheap source. The development of a bio-based surfactant from waste sources like waste cooking oil and non-edible jatropha oil has been reported [50,51]. The implementation of nanoparticles derived from fly ash has the potential to recover residual oil [52]. The derived carbon tube and synthesizing graphene from waste palm kernel can also be used for EOR applications. However, the above chemicals need further investigation and detailed analysis before they can be implemented for successful pilot or field projects.

Advanced Future Chemical EOR: Advanced chemical EOR schemes like nanofluid and ionic liquid flooding have been successfully investigated at a laboratory scale. However, detailed information is required regarding their adsorption, retention, porosity and permeability deterioration, and toxicity influence on the environment. Therefore, to commercialize the processes, intense research is required which can widen the scope of the current research field. Furthermore, the process of recycling the chemicals may have a huge impact on the process economy and simultaneously reduces environmental hazards. Hence, some experiments with the recycled chemicals must be performed to develop their feasibility for recycling operations. Additionally, the developed membrane to separate oil and chemicals from the produced water may not be effective to isolate nanoparticles or ionic liquid. Therefore for such a case, a new advanced membrane has to be developed [9].

6.5 Chemical EOR Laboratory and Pilot-Scale Studies

In the early 1980s, chemical EOR schemes in the United States have not been technically or economically feasible [53]. However, in recent years, successful chemical EOR has been conducted in various parts of the world. Alkali flooding was implemented in the field of Whittier (California, United States), Tpexozephoe (Russia), North Gujarat (India), Court Bakken (Saskatchewan, Canada) and Nagylengyel (Hungary) [54,55]. Surfactant flooding was reported in the field of Yates and Cottonwood Creek (Texas and Wyoming, United States), Mauddud (Bahrain) and Semoga (Indonesia). Polymer flooding was reported in the Tambaredjo field (Suriname), Mooney, Seal, Suffield Caen, Pelican Lake and East Bodo field (Canada), Vacuum field (Mexico), Marmul field (Oman), El Corcobo field (Argentina), West Cat Canyon and Albrecht field (Unites States) and Bohai Bay field (China) [55,56]. The combined alkali–surfactant–polymer (ASP) flooding impact was successfully carried out in different countries like China (Shengli, Daqing, Karamay, Jilin Hong-gang, Zhong-yuan Hu-zhuang-ji, Yumen Lao-jun-miao and Xinjiang oil fields), India (Viraj and Jhalora), United States (Cambridge Minnelusa, West Kiehl, Mellot Ranch, Sho-Vel-Tum, Lawrence, Tanner and Brookshire Dome), Venezuela (Lagomar) and Canada (Little Bow, Suffield, Taber South and Mooney) [1,3,55,57]. Whatsoever, the execution of chemical nanofluid EOR in any of the oil field has not been reported so far and extensive studies are in progress.

In this section, the applications of combined alkali–surfactant–polymer flooding on laboratory, pilot and commercial scales are discussed. A review article reported around 32 alkali–surfactant–polymer (ASP) flooding projects world-wide which have been conducted on pilot and large scale [3,26]. The majority of the tests were conducted in China (19 projects) followed by USA (7 projects), Canada (3 projects), India (2 projects), and Venezuela (1 project) as illustrated in Table 6.2. ASP flooding conducted in various fields like Elk Hills (California, USA), Mooney and Instow (Canada) and Caracara Sur (Columbia) lacks data and detailed information [3]. It was also reported that no ASP flooding was active in carbonate reservoirs; however, some SP flooding projects were reported [3]. Therefore based on the available data, some of the important pilot and field ASP flooding schemes have been formulated in Table 6.3.

TABLE 6.2

Overall Alkali-Surfactant-Polymer Flooding Test Worldwide [3]

Sr. No.	Country	Oil Field	Year of Initiation	References
1	China	Shengli Gudong	August, 1992	[58–60]
2		Yumen Lao-jun-miao	March, 1994	[61]
3		Daqing_Sa-zhong-xi	September, 1994	[58,62,63]
4		Daqing Xing-wu-zhong	January, 1995	[20,58,64]
5		Karamay	July, 1996	[64–68]
6		Daqing Xing-2-xi	September, 1996	[69]
7		Shengli Gudao-xi	May, 1997	[66,70,71]
8		Jilin Hong-gang	September, 1997	[72]
9		Daqing Sa-bei-1-xi	December, 1997	[73]
10		Zhong-yuan Hu-zhuang-ji	January, 2000	[74]
11		Daqing Xing-bei xing-Daqing	April, 2000	[20,75]
12		Daqing Sabei-bei-2-dong	October, 2004	[76]
13		DQ South-5	July, 2005	[77]
14		DQ North-1 East	June, 2006	[77]
15		DQ Xin 1-2	August, 2007	[77]
16		DQ North-2 West	November, 2008	[77]
17		DQ South-6	January, 2009	[77]
18		DQ Xin 6-East I	June, 2009	[77]
19		DQ Xin 6-East II	October, 2009	[77]
20	Unites States	West Kiehl	December, 1987	[78,79]
21		Cambridge	February, 1993	[80]
22		Tanner	May, 2000	[81]
23		Mellot Ranch	August, 2000	[82]
24		Lawrence	August, 2010	[83,84]
25		Brookshire Dome	September, 2011	[85]
26		Sho-Vel-Tum	-	[86]
27	Canada	Taber South	May, 2006	[87]
28		Mooney	2011	[88]
29		Little Bow	January, 2014	[89]
30	India	Viraj	August, 2002	[90]
31		Jhalora	February, 2010	[91]
32	Venezuela	Lagomar	-	[92,93]

6.5.1 ASP Flooding in China

In China, the ASP flooding scheme has been performed in Shengli, Daqing, Karamay, Xinjiang and Henan oil fields. Shengli and Daqing oil field are of sandstone type whereas Xinjiang and Karamay are conglomerate (heterogeneous) in nature [29]. The ASP flooding introduced in Shengli in the 1980s with the first test initiated in 1992 in the Gudong oil field that produces an additional oil recovery of 26% OOIP (original oil in place) [94]. The next pilot

TABLE 6.3

Summary of Chemical (Alkali-Surfactant-Polymer) EOR Pilot and Field Projects [1,77]

Country	Oil Field	Scale	Pattern	Year of Injection	Oil Category	Permeability	Oil Recovery
China	Daqing – west part xing-2 zone	Pilot	5 spots (4I, 9P)	-	-	0.858 μm^2	19.30% OOIP
	Daqing – west part of middle zone	Pilot	5 spots (4I, 9P)	-	-	0.809 μm^2	21.4% OOIP
	Daqing – middle part xing-5 zone	Pilot	5 spots (1I, 4P)	-	-	0.789 μm^2	25.0% OOIP
	Daqing - north middle zone	Pilot	4 spots (3I, 4P)	-	-	0.767 μm^2	23.24% OOIP
	Daqing – west part north-1 zone	Pilot	5 spots (6I, 12P)	-	-	0.812 μm^2	20.63% OOIP
	Karamay oil field – north part middle zone	Pilot	5 spots (4I, 9P)	-	-	0.157 μm^2	24.5% OOIP
	Daqing, north 1- east zone	Field	5 spots (49I, 63P)	2006	-	0.670 μm^2	22.5% OOIP predicted
	Daqing, north 2- west zone (weak alkali – ASP)	Field	5 spots (35I, 44P)	2008	-	0.533 μm^2	19.3% OOIP predicted
	Daqing – south-5 zone	Field	5 spots (29I, 39P)	2005	-	0.867 μm^2	18.6% OOIP predicted
	Daqing – middle part xing-2 zone	Field	5 spots (17I, 27P)	2000	-	0.85 μm^2	18.0% OOIP predicted
Unites States	Tanner	Pilot	-	-	-	-	17% OOIP
	Cambridge Minnelusa	Pilot	-	-	-	-	28% OOIP
	Sho-Vel-Tum	Pilot		1998			10,444 barrel of additional oil in 1.3 years
Canada	Taber south, Alberta	Pilot	5 spot	-	Medium gravity oil	-	Oil cut increases from 1.7% (300 bbl/day) to 7.3% (1502 bbl/day)
	Taber Glauconitic	-		-	-	-	Oil cut increases from 0.9% (26 m^3/dm) to 3.3% (127 m^3/dm)
India	Viraj	Pilot	-	2002	Heavy crude oil	4.5–9.9 Darcy	Oil recovery increased from 24.4 to 98.23 m^3/dm
	Jhalora	Pilot	-	2010	Heavy crude oil	1.9–8.7 Darcy	47,000 bbls additional oil recovery

test was conducted in Gudao reservoirs with 6 injection and 10 producer wells pattern resulting in an enhanced oil recovery of 15.5% OOIP. The reservoir average permeability and porosity were 1520 mD and 32% respectively. In the Daqing oil field, the ASP scheme was introduced in the year 1994 [62]. The pilot project has 4 injection and 5 production wells covering an area of 90,000 m^2. The formation is of sandstone type with average porosity and permeability of 26% and 1.426 μm^2 respectively. The performed ASP flooding showed a decrease in the water from 82.7% to 59.7%, which enhances the oil recovery to 91.5 from 36.7 m^3/d. The study on ASP flooding in central Xing 2 of the Daqing oil field predicted an additional oil recovery of above 20% by a numerical method and the original oil recovered (pilot test) was better than predicted [95]. The ASP in Karamay (conglomerate) heterogeneous reservoirs oil field was initiated in the year 1995 in five-spot patterns with a well spacing of 164 ft. The project was successful in decreasing the water cut that results in an enhanced oil recovery of around 25% OOIP [68]. Four field tests were conducted in the Daqing oil field with strong alkali (NaOH) and weak alkali (Na_2CO_3) ASP formulation as shown in Table 6.3. The three strong alkali ASP flooding predicted improved oil recoveries of 18%, 18.6% and 22.5% OOIP, respectively, whereas with weak alkali ASP flooding, the recovery predicted was around 19.3% OOIP [77].

The actual ASP flooding on an industrial scale was started in 2014 in the Daqing oil field China. After injecting ASP slug, the oil recovery was enhanced by 9% (3.5 million ton) and 11% (4 million ton) in the year 2015 and 2016 respectively. The ASP flooding was effective for a higher temperature of 81°C and resulted in an additional oil recovery of 7.7% [29].

6.5.2 ASP Flooding in the United States

In the United States, the ASP was introduced in West Khiel [78], Tanner [81], Cambridge Minnelusa [80], Lawrence field Illinois [96] and Sho-Vel-Tum [97] fields. In the year 1987, the initial pilot-scale ASP flooding was introduced in the field of West Kiehl, Wyoming [78,98]. The oil recovery in such a field was enhanced economically by 26% original oil in place. The Tanner and Cambridge Minnelusa field pilot project resulted in additional oil recovery of 17% and 28% OOIP with ASP flooding [81]. The core flooding test conducted in Cypress and Bridgeport sandstone (Lawrence field Illinois) enhances the oil recovery factor by 21% and 24% OOIP respectively. Finally, in the Sho-Vel-Tum field, the oil production was enhanced from 4 to 20 bbl/day and this pilot project started in 1998 supplements another 10,444 barrel of oil in 1.3 years.

6.5.3 ASP Flooding in Canada

The oil reservoirs of Canada in which ASP is executed are Taber South, Taber Glauconitic and Suffield [1,99]. In 2006, the project was performed in Taber South where the oil production increases from 300 to 1502 bbl/day which accounts for an oil cut of 7.3% from 1.7%. The project also indicated that part of the oil reservoirs is active for ASP flooding and is dependent on the reservoir properties. The ASP flooding in the Taber Glauconitic field continued from 2006 to 2010 and during this period, the oil cut was enhanced from 0.9% (26 m^3/dm) to 5.3% (127 m^3/dm). In Suffield, the main objectives of ASP flooding were to minimize the quantity of fresh water required for chemical slug and the reusability of the produced water. The pilot project was successful in terms of its objectives and was heading for large-scale operations.

6.5.4 ASP Flooding in India

The ASP flooding in India was performed in two fields named Viraj and Jhalora both located in Ahmedabad, Cambay basin of India [90,91]. The ASP flooding in the Viraj oil field was injected in August 2002, which showed an increment in oil recovery from 24.4 to 98.23 m^3/dm which reduces the water cut from 83.5% to 71.4% (time period from August 2002 to March 2003) [90]. In Jhalora oil fields, when an ASP slug of 0.17 PV was injected, an additional oil of around 47,000 bbls was achieved resulting in an encouraging outcome [91].

References

1. A. A. Olajire, Review of ASP EOR (alkaline surfactant polymer enhanced oil recovery) technology in the petroleum industry: Prospects and challenges, *Energy*, vol. 77, pp. 963–982, 2014.
2. M. A. Bataweel and H. A. Nasr-El-Din, Alternatives to minimize scale precipitation in carbonate cores caused by alkalis in asp flooding in high salinity/high temperature applications, in *Proceeding of the SPE European Formation Damage Conference*, Noordwijk, The Netherlands, SPE 143155, 2011, pp. 1–11.
3. J. J. Sheng, A comprehensive review of alkaline-surfactant-polymer (ASP) flooding, *Asia-Pacific Journal of Chemical Engineering*, vol. 9, pp. 471–489, 2014.
4. A. Weatherill, Surface development aspects of alkali-surfactant-polymer (ASP) flooding, *Presented at the International Petroleum Technology Conference*, Doha, Qatar, 2009.
5. J. J. Sheng, *Modern Chemical Enhanced Oil Recovery Theory and Practice*. Gulf Professional Publishing, Elsevier, Houston, TX, 2011.
6. S. Bridgewater, A. Goldszal, and A. Milani, Polymer EOR: A new HSE hazard in an old industry, *Presented at the International Petroleum Technology Conference, IPTC-18659-MS*, Bangkok, Thailand, 2016.
7. H. J. Hill and L. W. Lake, Cation exchange in chemical flooding: Part 3 e experimental, *Society of Petroleum Engineers Journal*, vol. 18, pp. 445–456, 1978.
8. F. Civan (Ed.), Chapter 1: Overview of formation damage, in *Reservoir Formation Damage*, 3rd edn. Gulf Professional Publishing, Boston, MA, pp. 1–6, 2016.

9. A. O. Gbadamosi, R. Junin, M. A. Manan, A. Agi, and A. S. Yusuff, An overview of chemical enhanced oil recovery: Recent advances and prospects, *International Nano Letters*, vol. 9, pp. 171–202, 2019.
10. B. Yuan and D. A. Wood, Chapter 1: Overview of formation damage during improved and enhanced oil recovery, in *Formation Damage during Improved Oil Recovery*. Gulf Professional Publishing, Houston, TX, pp. 1–20, 2018.
11. M. I. Youssif, R. M. El-Maghraby, S. M. Saleh, and A. Elgibaly, Silica nanofluid flooding for enhanced oil recovery in sandstone rocks, *Egyptian Journal of Petroleum*, vol. 27, pp. 105–110, 2018.
12. T. Sharma, S. Iglauer, and J. S. Sangwai, Silica nanofluids in an oilfield polymer polyacrylamide: Interfacial properties, wettability alteration, and applications for chemical enhanced oil recovery, *Industrial & Engineering Chemistry Research*, vol. 55, pp. 12387–12397, 2016.
13. H. Ehtesabi, M. M. Ahadian, V. Taghikhani, and M. H. Ghazanfari, Enhanced heavy oil recovery in sandstone cores using TiO_2 nanofluids, *Energy and Fuels*, vol. 28, pp. 423–430, 2014.
14. R. Saha, R. Uppaluri, and P. Tiwari, Impact of natural surfactant (Reetha), polymer (Xanthan Gum), and silica nanoparticles to enhance heavy crude oil recovery, *Energy and Fuels*, vol. 33, pp. 4225–4236, 2019.
15. R. Saha, R. Uppaluri, and P. Tiwari, Silica nanoparticle assisted polymer flooding of heavy crude oil: Emulsification, rheology, and wettability alteration characteristics, *Industrial and Engineering Chemistry Research*, vol. 57, pp. 6364–6376, 2018.
16. D. T. Nguyen and N. Sadeghi, Stable emulsion and demulsification in chemical EOR flooding: Challenges and best practices, *Presented at the SPE EOR Conference Oil Gas West Asia*, Muscat, Oman, 2012.
17. M. Li, J. Guo, B. Peng, M. Lin, Z. Dong, and Z. Wu, Chapter 14: Formation of crude oil emulsions in chemical flooding, in Sjoblom, J, (Ed.) *Emulsions and Emulsion Stability*, 2nd edn. CRC Press, Boca Raton, FL, 2006, pp. 517-547.
18. H. B. Li, *Advances in Alkaline-Surfactant-Polymer Flooding and Pilot Tests*. Science Press, China, 2007.
19. W. L. Kang and D. M. Wang, Emulsification characteristic and de-emulsifiers action for alkaline/surfactant/polymerflooding, in *SPE Asia Pacific Improved Oil Recovery Conference, SPE 72138*, Kuala Lumpur, Malaysia, 2001.
20. H. Z. Wang, G. Z. Liao, and J. Song. (2006). Combined chemical flooding technologies.
21. D. Li, M. Shi, D. Wang, and Z. Li, Chromatographic separation of chemicals in alkaline Surfactant Polymer Flooding in Reservoir Rocks in the Daqing Oil Field, in *SPE International Symposium on Oilfield Chemistry*, The Woodlands, Texas, 2009.
22. D. M. Wang, Development of new tertiary recovery theories and technologies to sustain Daqing oil production, *Petroleum Geology & Oilfield Development in Daqing*, vol. 20, pp. 1–7, 2001.
23. W. Demin, J. Youlin, W. Yan, G. Xiaohong, and W. Gang, Viscous-elastic polymer fluids rheology and its effect upon production equipment, *SPE Production & Facilities*, vol. 19, pp. 209–216, 2004.
24. K. Raney, S. Ayilara, R. Chin, and P. Verbeek, Surface and subsurface requirements for successful implementation of offshore chemical enhanced oil recovery, in *Offshore Technology Conference, SPE Paper OTC 21188*, Houston, TX, 2011.

25. D. Morel, M. Vert, S. Jouenne, R. Gauchet, and Y. Bouger, First polymer injection in deep offshore field, in *SPE Paper No. 135735*, Angola: recent advances on Dalia/Camelia field case, 2010.
26. H. Guo, Y. Li, F. Wang, Z. Yu, Z. Chen, Y. Wang, et al., ASP flooding: Theory and practice progress in China, *Journal of Chemistry*, vol. 2017, pp. 1–18, 2017. doi: 10.1155/2017/8509563.
27. D. Denney, Pump-scaling issues in ASP flooding in Daqing oil field, SPE-0108–0050-JPT, *Journal of Petroleum Technology*, vol. 6, pp. 50–52, 2008. doi: 10.2118/0108-0050-JPT.
28. C. Jiecheng, Z. Wanfu, Z. Yusheng, G. Xu, C. Ren, P. Zhangang, et al., Scaling principle and scaling prediction in ASP flooding producers in Daqing oilfield, SPE-144826-MS, in *SPE Enhanced Oil Recovery Conference*, 19–21 July, Kuala Lumpur, Malaysia, 2011. doi: 10.2118/144826-MS.
29. H. Guo, Y. Li, Y. Li, D. Kong, B. Li, and F. Wang, Lessons learned from ASP flooding tests in China, SPE-186036-MS, in *SPE Reservoir Characterisation and Simulation Conference and Exhibition*, 8–10 May, Abu Dhabi, 2017. doi: 10.2118/186036-MS.
30. H. Guo, Y. Q. Li, Y. Zhu, F. Y. Wang, D. B. Kong, and R. C. Ma, Comparison of scaling in strong alkali and weak alkali ASP flooding pilot tests, in *Conference Proceedings, IOR 2017: 19th European Symposium on Improved Oil Recovery*, April 2017, pp. 1–13. doi: 10.3997/2214-4609.201700258.
31. H. Guo, Y. Li, F. Wang, and Y. Gu, Comparison of strong-alkali and weak-alkali ASP-flooding field tests in Daqing oil field, SPE-179661, *SPE Production & Operations*, vol. 33, pp. 1–10, 2018. doi: 10.2118/179661-PA.
32. H. ShamsiJazeyi, R. Verduzco, and G. J. Hirasaki, Reducing adsorption of anionic surfactant for enhanced oil recovery: Part I. Competitive adsorption mechanism, *Colloids and Surfaces A: Physicochemical and Engineering Aspects*, vol. 453, pp. 162–167, 2014.
33. G. Kalfoglou, Lignosulfonates as sacrificial agents in oil recovery processes, US Patent No. 4006779, 1977.
34. J. Novosad, Laboratory evaluation of lignosulfonates as sacrificial adsorbates in surfactant flooding, *Journal of Canadian Petroleum Technology*, vol. 23, pp. 1–6, 1984.
35. J. Qing, B. Zhou, R. Zhang, Z. Chen, and Y. Zhou, Development and application of a silicate scale inhibitor for asp flooding production scale, in *International Symposium on Oilfield Scale*, Aberdeen, United Kingdom, 2002, pp. 1–4.
36. D. Conne, Prediction and treatment of scale in North Sea fields, M.E. Thesis, Heriot-Watt University, 1983.
37. M. Crabtree, E. Eslinger, P. Fletcher, A. Johnson, and G. King, Fighting scale: Removal and prevention, *Oilfield Review*, vol. 11, pp. 30–45, 1999.
38. C. Bezemer and A. K. Bauer, Prevention of carbonate scale deposition: A well-packing technique with controlled solubility phosphates, *Journal of Petroleum Technology*, vol. 21, pp. 505–514, 1969.
39. T. Chen, A. Neville, and M. Yuan, Effect of PPCA and DETMPP inhibitor blends on $CaCO_3$ scale formation, in *Proceeding of the 6th International Symposium on Oil Field Scale, SPE 87442*, Aberdeen, United Kingdom, 2004, pp. 1–7.
40. S. Taj, S. Papavinasam, and R. W. Revie, Development of green inhibitors for oil and gas applications, in *Proceeding of the CORROSION*, San Diego, California, 2006, pp. 1–9.

41. W. W. Frenier and D. Wilson, Use of highly acid-soluble chelating agents in well stimulation services, in *Proceeding of the SPE Annual Technical Conference and Exhibition, SPE 63242,* Dallas, Texas, 2000, pp. 1–12.
42. T. Kumar, S. Vishwanatham, and S. S. Kundu, A laboratory study on pteroyl-L-glutamic acid as a scale prevention inhibitor of calciumcarbonate in aqueous solution of synthetic produced water, *Journal of Petroleum Science & Engineering,* vol. 71, pp. 1–7, 2010.
43. N. Kohler, B. Bazin, A. Zaitoun, and T. Johnson, Green inhibitors for squeeze treatments: A promising alternative, in *Proceeding of the CORROSION, Paper 04537,* New Orleans, LA, NACE, 2004, pp. 1–19.
44. K. A. Elraies, The effect of a new in situ precipitation inhibitor on chemical EOR, *Journal of Petroleum Exploration and Production Technology,* vol. 3, pp. 133–137, 2013.
45. Z. Amjad, Effect of precipitation inhibitors on calcium phosphate scale formation, *Canadian Journal of Chemistry,* vol. 67, pp. 850–856, 1989.
46. K. A. Elraies and I. M. Tan, Design and application of a new acide alkalie surfactant flooding formulation for Malaysian reservoirs, in *SPE Asia Pacific Oil and Gas Conference and Exhibition, SPE 133005,* Brisbane, Australia, 2010.
47. M. Algharaib, A. Alajmi, and R. Gharbi, Improving polymer flood performance in high salinity reservoirs, *Journal of Petroleum Science & Engineering,* vol. 115, pp. 17–23, 2014.
48. M. K. Algharaib and M. A. Abedi, Optimization of polymer flood performance by preflush injection: numerical investigation, in *SPE Kuwait International Petroleum Conference and Exhibition,* Kuwait City, Kuwait, 2012, pp. 1–14.
49. C. Sanz and G. Pope, Alcohol-free chemical flooding: from surfactant screening to coreflood design, in *SPE International Symposium on Oilfield Chemistry, SPE Paper No. 28956,* San Antonio, Texas, 1995.
50. S. Kumar, N. Saxena, and A. Mandal, Synthesis and evaluation of physico-chemical properties of anionic polymeric surfactant derived from Jatropha oil for application in enhanced oil recovery, *Journal of Industrial and Engineering Chemistry,* vol. 43, pp. 106–116, 2016.
51. Q. Q. Zhang, B. X. Cai, W. J. Xu, H. Z. Gang, J. F. Liu, S. Z. Yang, et al., The rebirth of waste cooking oil to novel bio- based surfactants, *Scientific Report,* vol. 5, p. 9971, 2015.
52. A. A. Eftekhari, R. Krastev, and R. Farajzadeh, Foam stabilized by fly ash nanoparticles for enhancing oil recovery, *Industrial and Engineering Chemistry Research,* vol. 54, pp. 12482–12491, 2015.
53. L. N. Nwidee, S. Theophilus, A. Barifcani, M. Sarmadivaleh, and S. Iglauer, EOR processes, opportunities and technological advancements. in Romero-Zeron, L. (Ed.), *Chemical Enhanced Oil Recovery (cEOR) – a Practical Overview.* InTech, pp. 1–50, 2016, doi: 10.5772/64828.
54. S. Kumar, T. F. Yen, G. V. Chilingarian, and E. C. Donaldson, Chapter 9 Alkaline flooding, in Donaldson, E.C., Chilingarian, G.V., Yen, T.F., (Eds.), Developments in Petroleum Science, Elsevier, vol. 17 (B), pp. 219–254, 1989, doi: 10.1016/S0376-7361(08)70461-8.
55. J. J. Sheng, Enhanced oil recovery field case studies, Gulf Professional Publishing, pp. 1–712, 2013.
56. H. Saboorian-Jooybari, M. Dejam, and Z. Chen, Heavy oil polymer flooding from laboratory core floods to pilot tests and field applications: Half-century studies, *Journal of Petroleum Science & Engineering,* vol. 142, pp. 85–100, 2016.

57. T. French, Evaluation of the Sho-Vel-Tum Alkali-Surfactant-Polymer (ASP) oil recovery project, *DOE/SW/45030-1; U.S. Department of Energy,* Tulsa, Oklahoma, 1999.
58. D. M. Wang, Z. H. Zhang, J. C. Cheng, J. C. Yang, S. T. Gao, and L. Li, Pilot tests of alkaline/surfactant/polymer flooding in Daqing oil field; spe reservoir engineering, vol. 12, pp. 229–233, 1997.
59. Z. J. Qu, Y. G. Zhang, X. S. Zhang, and D. J. L, A successful ASP flooding pilot in Gudong oil field, SPE/DOE 39613, in *Improved Oil Recovery Symposium,* 19–22 April, Tulsa, Oklahoma, 1998.
60. W. C. Song, C. Z. Yang, D. K. Han, Z. J. Qu, B. Y. Wang, and W. L. Jia, Alkaline–surfactant–polymer combination flooding for improving recovery of the oil with high acid value, SPE-29905, in *International Meeting on Petroleum Engineering,* 14–17 November, Beijing, China, 1995.
61. D. C. Wang, T. R. Yang, T. P. Du, B. F. Fang, and C. Z. Yang, Micellar/polymer flooding pilot test in the H184 well pattern in the Laojunmiao field, *Petroleum Exploration and Development,* vol. 26, pp. 47–49, 1999.
62. G. Shutang, L. Huabin, Y. Zhenyu, M. Pitts, H. Surkalo, and K. Wyatt, Alkaline/surfactant/polymer pilot performance of the west central Saertu, Daqing oil field, *SPE Reservoir Engineering,* vol. 11, pp. 181–188, 1996.
63. H. B. Li, H. F. Li, Z. Y. Yang, Z. G. Ye, and J. C. Cheng, ASP pilot test in Daqing Sa-Zhong-Xi region, *Oil and Gas Recovery Technology,* vol. 6, pp. 15–19, 1999.
64. D. Han, *Surfactant Flooding: Principles and Applications,* Petroleum Industry Press, Beijing, China, 2001.
65. M. Delshad, W. Han, G. A. Pope, K. Sepehrnoori, W. Wu, R. Yang, et al., Alkaline/surfactant/polymer flood predictions for the Karamay oil field, SPE-39610, in *SPE/DOE Improved Oil Recovery Symposium,* 19–22 April, Tulsa, Oklahoma, 1998.
66. H. L. Chang, Z. Q. Zhang, Q. M. Wang, Z. S. Xu, Z. D. Guo, H. Q. Sun, et al., Advances in polymer flooding and alkaline/surfactant/polymer processes as developed and applied in the People's Republic of China, *Journal of Petroleum Technology,* vol. 58, pp. 84–89, 2006.
67. H. J. Gu, R. Q. Yang, S. G. Guo, W. D. Guan, X. J. Yue, and Q. Y. Pan, Study on reservoir engineering: ASP flooding pilot test in Karamay oilfield, SPE-50918, in *International Oil and Gas Conference and Exhibition,* 2–6 November, Beijing, China, 1998.
68. Q. Qiao, H. Gu, D. Li, and L. Dong, The pilot test of ASP combination flooding in Karamay oil field, SPE-64726, in *SPE International Oil and Gas Conference and Exhibition in China,* 7–10 November, Beijing, China, 2000. doi: 10.2118/64726-MS.
69. D. Wang, J. Cheng, J. Wu, F. Wang, H. Li, and X. Gong, An alkaline/surfactant/polymer field test in a reservoir with a long-term 100% water cut, SPE-49018-MS, in *SPE Annual Technical Conference and Exhibition,* 27–30 September, New Orleans, Louisiana, 1998.
70. Z. J. Yang, Y. Zhu, X. Y. Ma, X. H. Li, J. Fu, P. Jiang, et al., ASP pilot test in Guodao field, *Journal of Jianghan Petroleum Institute,* vol. 24, pp. 62–64, 2002.
71. X. L. Cao, H. Q. Sun, Y. B. Jiang, X. S. Zhang, and L. L. Guo, ASP pilot test in the Western Guodao field, *Oilfield Chemistry,* vol. 19, pp. 350–353, 2002.
72. Z. L. Zhang, Y. M. Yang, L. Hong, T. Peng, and M. R. Xuan, ASP pilot test in Honggang field, *Journal of Southwest Petroleum Institute,* vol. 23, pp. 47–57, 2001.

73. D. M. Wang, J. C. Cheng, Q. Li, L. Z. Li, C. J. Zhao, and J. C. Hong, An alkaline bio-surfactant polymer flooding pilots in Daqing Oil Field, SPE-57304, in *SPE Asia Pacific Improved Oil Recovery Conference*, 25–26 May, Kuala Lumpur, Malaysia, 1999.
74. J. L. Jiang, D. F. Guo, X. F. Li, P. W. Zhao, X. J. Wang, X. L. Wu, et al., Pilot field trial of natural mixed carboxylates/xanthan flood at well H5–15 block in Huzhuangji oil field, *Oilfield Chemistry*, vol. 20, pp. 58–60, 2003.
75. H. F. Li, G. Z. Liao, P. H. Han, Z. Y. Yang, X. L. Wu, G. Y. Chen, et al., Alkaline/surfactant/polymer (ASP) commercial flooding test in the central Xing2 area of Daqing Oilfield, SPE-84896, in *International Improved Oil Recovery Conference in Asia Pacific*, 20–21 October, Kuala Lumpur, Malaysia, 2003.
76. X. Wan, M. Huang, H. Yu, Y. Zhang, W. Chen, and E. Gao, Pilot test of weak alkaline system ASP flooding in secondary layer with small well spacing, SPE-104416-MS, in *International Oil & Gas Conference and Exhibition in China*, 5–7 December, Beijing, China, 2006. doi: 10.2118/104416-MS.
77. Y. Zhu, Q. Hou, W. Liu, D. Ma, and G. Liao, Recent progress and effects analysis of ASP flooding field test, SPE-151285, in *Eighteenth SPE Improved Oil Recovery Symposium*, 14–18 April, Tulsa, Oklahoma, 2012. doi: 10.2118/151285-MS.
78. J. J. Meyers, M. J. Pitts, and K. Wyatt, Alkaline-surfactant-polymer flood of the West Kiehl, Minnelusa unit, SPE-24144-MS, in *SPE/DOE Enhanced Oil Recovery Symposium*, 22–24 April, Tulsa, Oklahoma, 1992. doi: 10.2118/24144-MS.
79. S. R. Clark, M. J. Pitts, and S. M. Smith, Design and appication of an alkaline-surfactan-polymer recovery system to the West Kiehl field, *SPE Advanced Technology Series*, vol. 1, pp. 172–179, 1993.
80. J. Vargo, J. Turner, B. Vergnani, M. Pitts, K. Wyatt, H. Surkalo, et al., Alkaline-surfactant-polymer flooding of the Cambridge Minnelusa field, SPE-68285, *SPE Reservoir Evaluation & Engineering*, vol. 3, pp. 552–558, 2000.
81. M. J. Pitts, P. Dowling, K. Wyatt, H. Surkalo, and C. Adams, Alkaline-surfactante-polymer flood of the tanner field, in *SPE/DOE Symposium on Improved Oil Recovery*, 22–26 April, Tulsa, Oklahoma, 2006, pp. 1–5.
82. http://www.surtek.com/mellottranch.html (accessed on November 2, 2013).
83. A. Sharma, A. Azizi-Yarand, B. Clayton, G. Baker, P. McKinney, C. Britton, et al., The design and execution of an alkaline-surfactant-polymer pilot test, SPE-154318, in *SPE Improved Oil Recovery Symposium*, 14–18 April, Tulsa, Oklahoma, 2012. doi: 10.2118/154318-MS.
84. M. Dean, Selection and evaluation of surfactants for field pilots, M.S. Thesis, University of Texas at Austin, 2011.
85. C. Lewis, M. M. Sharma, and B. Gao, Field demonstration of alkaline surfactant polymer floods in mature oil reservoirs, in *Brookshire Dome*, Texas, RPSEA report 08123.02, 2012.
86. T. R. French, Evaluation of the Sho-Vel-Tum alkaline-surfactant-polymer (ASP) oil recovery project, Stephens County, OK, Contract No. DOE/SW/45030-1, OSTI ID: 9726, US DOE, Washington, DC, July, 1999.
87. L. E. McInnis, K. D. Hunter, T. T. Ellis-Toddington, and D. J. Grawbarger, Case study of the Taber Mannville B ASP flood, SPE-165264-MS, in *SPE Enhanced Oil Recovery Conference*, 2–4 July, Kuala Lumpur, Malaysia, 2013. doi: 10.2118/165264-MS.

88. S. H. Liu, D. L. Zhang, W. Yan, M. Puerto, G. J. Hirasaki, and C. A. Miller, Favorable attributes of alkaline-surfactant-polymer flooding, *Society of Petroleum Engineers Journal*, vol. 13, pp. 5–16, 2008.
89. www.zargon.ca (accessed on April 15, 2014).
90. M. Pratap and M. S. Gauma, Field implementation of alkaline-surfactant-polymer (ASP) flooding: A maiden effort in India, SPE-88455, in *SPE Asia Pacific Oil and Gas Conference and Exhibition*, Perth, Austrlia, 18–20 October, 2004, pp. 1–5.
91. A. K. Jain, A. K. Dhawan, and T. R. Misra, ASP flood pilot in Jhalora (K-IV): A case study, SPE-153667, in *SPE Oil and Gas India Conference and Exhibition*, 28–30 March, Mumbai, India, 2012. doi: 10.2118/153667-MS.
92. G. Manrique, G. D. Carvajal, L. Anselmi, C. Romero, and L. Chacon, Alkali/surfactant/polymer at VLA 6/9/21 field in Maracaibo lake: Experimental results and pilot project design, in *SPE/DOE Improved Oil Recovery Symposium*, 3–5 April, Tulsa, Oklahoma, 2000.
93. C. Hernandez, L. Chacon, L. Anselmi, R. Angulo, E. Manrique, E. Romero, et al., Single well chemical tracer test to determine ASP injection efficiency at Lagomar VLA-6/9/21 area, C4 member, lake Maracaibo, Venezuela, SPE-75122-MS, in *SPE/DOE Improved Oil Recovery Symposium*, 13–17 April, Tulsa, Oklahoma, 2002.
94. C. Wang, B. Wang, X. Cao, and H. Li, Application and design of alkaline-surfactant-polymer system to close well spacing pilot Gudong oilfield, SPE 38321, in *SPE Western Regional Meeting*, 25–27 June, Long Beach, California, 1997.
95. D. Wang, J. Cheng, J. Wu, Z. Yang, Y. Yao, and H. Li, Summary of ASP pilots in Daqing oil field, in *SPE Asia Pacific Improved Oil Recovery Conference*, 25–26 October, Kuala Lumpur, Malaysia, 1999.
96. B. Seyler, J. Grube, B. Huff, N. Webb, J. Damico, C. Blakley, et al., Reservoir characterization of Bridgeport and cypress sandstones in Lawrence field Illinois to improve petroleum recovery by alkaline-surfactant-polymer flood, *Energy Technology Data Exchange*, 2012. doi: 10.2172/1083760.
97. T. French, Evaluation of the Sho-Vel-Tum alkali-surfactant-polymer (ASP) oil recovery project, DOE/SW/45030-1, Tulsa, Oklahoma, U.S. Department of Energy, 1999.
98. S. Clark, M. Pitts, and S. Smith, Design and application of an alkaline-surfactant-polymer recovery system to the west Kiehl field, *SPE Advanced Technology Series*, vol. 1, pp. 172–179, 1993.
99. M. Charest, Alkaline-surfactant-polymer (ASP) flooding in Alberta: Small amounts of the right chemicals can make a big difference, *Canadian Discovery Digest*, vol. 1, pp. 20–52, 2013.

7

Application of Nanotechnology in Unconventional Reservoirs

7.1 Introduction

Crude oil production from conventional resources by primary, secondary and tertiary (EOR) techniques has already been covered in the earlier chapters. However, the oil and gas recovery from unconventional resources like shale oil and tight oil reservoirs is further challenging due to lower permeability, reduced porosity and extreme heterogeneity of the formation. The schematic diagram of permeability distribution for conventional and unconventional reservoirs is depicted in Figure 7.1 [1–3]. The unconventional resources have been invented several decades ago but the natural flow of oil and gas through such formation is restricted due to poor connectivity of the pores unlike conventional reservoirs [2,4,5]. The production of reservoir fluids from unconventional reserves can be improved by increasing the porosity and permeability of the formation. The primary production can be obtained by the combined effect of multistage fracturing and horizontal drilling where the economic feasibility of the process faces a severe issue [6–8]. The studies reported that an oil recovery factor of 5%–15% can be achieved for tight oil reservoirs possessing a median porosity of 20% and permeability in the range of micro-Darcy to milli-Darcy (μD to mD) whereas for shale oil, the recovery factor of oil was estimated around 1%–10% (porosity < 15% and permeability < 1 mD) [9,10]. Therefore, the amount of residual or trapped oil in the unconventional reservoir is large as compared to conventional reservoirs.

In the past couple of decades, the petroleum industries in North America have emerged as role models towards commercializing the oil and gas production schemes from unconventional resources. Natural gas production around three quarters and almost 50% of the total petroleum liquids products have been produced from shale oil and tight reservoirs in the United States as of 2015 [11]. The oil recovery data are expected to improve further in upcoming decades and this outcome along with the depletion condition of conventional reservoirs had made a tremendous impact on several countries like Argentina, China, etc. to explore unconventional resources.

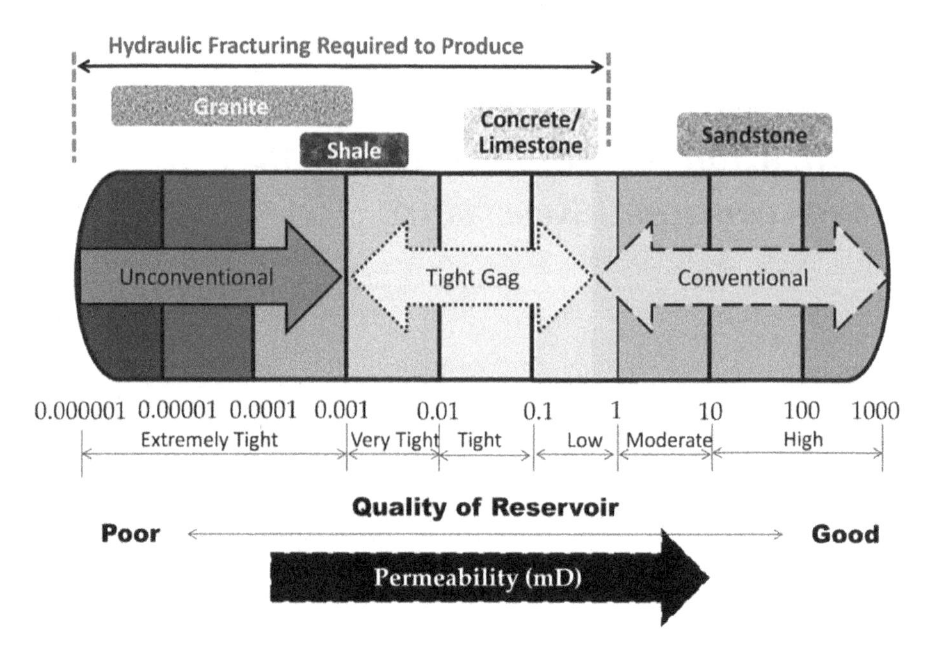

FIGURE 7.1
Permeability distribution of conventional and unconventional reservoirs.

Hydraulic fracturing stimulation is deployed to reservoirs with rock permeability <1 mD (Figure 7.1) [3,12]. Hydraulic fracturing is a process in which fracturing fluid along with proppants like ceramic or bauxite granules and sand is introduced in the formation rock at high pressure. The fracturing fluid when injected into the payzone initiates and propagates the fractures and further acts as a supporting medium for carrying and distributing the proppant in the fractures. Once the pressure is released, the openings initiated by the pressurized fracturing fluid tend to collapse, which is avoided by the presence of a proppant. The proppant is responsible for holding the openings of the fractures and thus maintains proper conductivity of the fractures [13]. Therefore, the permeability of the tight formation is enhanced and hydrocarbons can then flow easily towards the wellbore [3,14].

The major challenge to perform effective hydraulic fracturing in unconventional reservoirs is the development of fracturing fluid. The fracturing fluid should preserve outstanding rheology properties at high temperature and pressure with shear resistance as encountered under the downhole conditions.

Currently, researchers are actively working on the concept of hydraulic fracturing stimulation by the application of nanotechnology [15–17]. The desired properties of nanoparticles such as their size (1–100 nm), high surface area, and stability at higher temperature and higher pressure (high shear rate) can effectively enhance their applications under downhole conditions [18,19]. The required thermal, electrical, mechanical and optical properties for applications in unconventional reservoirs can be achieved by modifying the physical and chemical properties of the nanoparticles [20,21].

Conventional fracturing fluids along with their limitations are discussed in this chapter. Furthermore, the measurements to improve the rheological properties of fracturing fluid for its application under downhole conditions are elaborated. Finally, a review on the applications of nanotechnology in unconventional reservoirs with its challenges and limitations is included to identify the lacuna for further studies.

7.2 Hydraulic Fracturing Fluid

The fluid employed for hydraulic fracturing includes a base fluid, proppant and some chemical additives such as friction reducers, breakers, clay stabilizers, biocides, fluid loss control agent, etc [3,22,23]. The most deploying fracturing fluids are water-based fluids, polymer-based fluids, cross-linked fluids, viscoelastic surfactant (VES) fluids and foam-based fluids [3,17,22,24]. In water-based fluids, the fluid contains only water and proppants known as slick-water fluid, while polymer-based fluid consists of water, proppant and gelling agent known as a linear polymeric fluid. The fluids which comprise water, proppant, gelling agent and cross-linker are defined as cross-linked fluids. Viscoelastic surfactant (VES) fluids mainly involve water, proppant and surfactants, whereas foam and emulsion-based fluids include surfactants, polymers and nanoparticles as a stabilization/foaming agent along with gas dispersed in a liquid.

Water-based fluid is usually chosen over oil-based fluid because of its eco-friendly nature and economic consideration [25,26]. Water-based fluids such as slick-water possess low viscosity and has the ability to initiate less wide but longer fractures without any unnecessary increase in the height of fracture [13,27]. The complex formations which are fragile and naturally fractured along with the ability to sustain a large quantity of water are appropriate for slick-water [25,27]. Moreover, they are usually employed for shale gas reservoirs in which the condition of higher hydraulic fracturing conductivity is not essential [25].

Polymer-based fluids are those fluids in which biopolymers (guar and cellulose derived) and synthetic polymers (polyacrylamide) are added to enhance the viscosity of the fracturing fluid. Biopolymers that are guar based include hydroxyl propyl guar (HPG) and carboxylmethyl hydroxyl propyl guar (CMHPG) and cellulose-based polymers include carboxymethyl cellulose (CMC) and carboxymethyl hydroxyl ethyl cellulose (CMHEC) [3,24,28]. Similarly, the examples of polyacrylamide-based (synthetic) polymers are PHPA (partially hydrolysed polyacrylamide), AMPS (2-acrylamido-2-methylpropanesulfonic acid) and PVA (Vinyl phosphonate) [24,29–31].

The biopolymer-based fracturing fluid was first used for acid fracturing treatment in the year 1953 [32]. Guar-based polymer is most widely preferred to perform fracturing because of its low cost [3,33]. Guar gum polymer possesses the highest viscosity with respect to other available natural gums, minimum swelling rate and degradables by the attack of micro-organisms [34]. Studies reported that guar gum retains shear thinning behaviour which is desirable for hydraulic fracturing operations [35,36]. The polymer developed a layer of filter cake at the fracture faces, which prevents the leak-off of the fluid in the formation [37]. Guar-based polymers are not suitable for high temperature and hence to develop fracturing fluid under such conditions, synthetic polymers have usually been opted [24]. The desired properties of the polymeric-based fluid are that they should be resistant against high shear rate, minimum frictional loss, enough viscosity to transport proppant, minimum leak-off into the formation matrix, minimum formation damage, compatibility and clean-up efficiency [24]. They are economical, simple in operation, highly soluble, biodegradable, non-toxic and environmentally friendly [21,38].

Cross-linked polymer-based fluids are generally employed for optimum performance under harsh downhole conditions instead of excess polymer loading. It includes the combination of polymeric gel fluid with cross-linkers like metallic (zirconium and titanium) cross-linkers and borate esters that assist in enhancing the required viscosity [3,24,39].

The viscoelastic surfactant fluid is a mixture of surfactants and co-surfactants with inorganic salts to enhance the viscous and elastic properties. The mixture results in an orderly structure (3D network) categorized as worm-like or rod-like micelles and vesicles or lamellar structures as shown in Figure 7.2 [3,40,41]. The arrangement of these structures depends on the

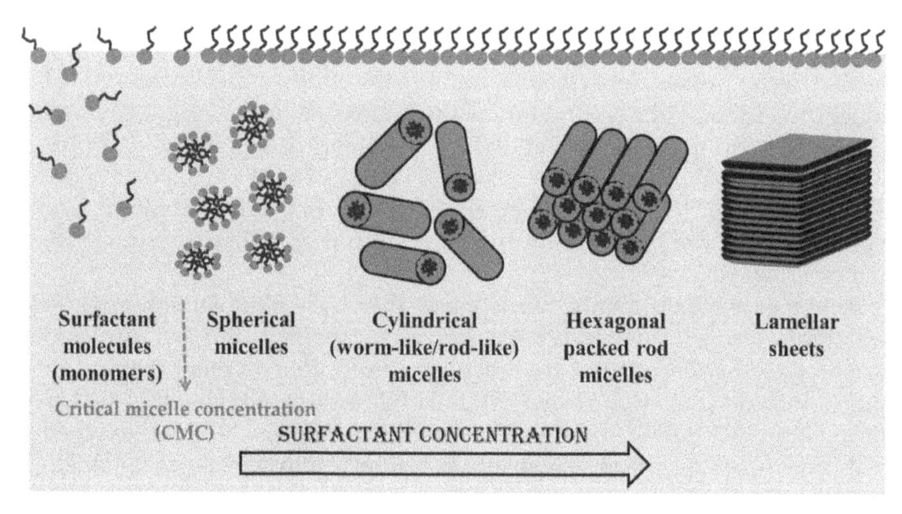

FIGURE 7.2
Formation of different surfactant structures/micelles in the VES solution.

interaction between the surfactant molecules (their shape and size) when the concentration is used above CMC and is amplified by inducing electrolytes. The surfactants engaged in the mixture are cationic–anionic, anionic–anionic, zwitterionic–amphoteric and zwitterionic–ionic [42–48]. The advantages of VES fluid are being water-soluble, with lower surface tension, low friction and breakdown of velocity after fracturing and no residue. They are simple in operation, suitable for different formation water and oil, require lesser energy as compared to polymer-based fluid and has no requirement for any chemical breaker/biocides/clay control additives [33,40,48–52].

Foamed fluid is a mixture of gas–liquid solution widely used as fracturing fluid media for water-sensitive high clay content unconventional reservoirs. The most used gases to produce foam are CO_2 and N_2 and they are friendly to the environment and formation [3,24]. Foam-based fluid can minimize issues like channelling, fingering and bypassing. It has the advantages of proper transport and placement of the proppant, less leak-off and an effective clean-up process [13,53–56].

7.3 Limitations of Hydraulic Fracturing Fluid

Slick water fluid due to lower apparent velocity reduces the proppants carrying capacity [57], disturbs the desired suspension and transportation of the proppants at the fractures and has no fluid loss control properties [22]. It has been reported that the water-based fluid, possessing low viscosity of 2–3 cp, can transport the proppant particles up to the size of <400 µm [58]. For optimum transport of the proppant, the fracturing fluid should possess a higher viscosity of above 100 cp at a shear rate of 100 per seconds [59]. Moreover, the majority of the unconventional reservoirs are water-sensitive in nature, which results in clay swelling, therefore reducing the relative permeability of oil and gas. Therefore, hydrocarbon recovery by fracturing from such formation is not suitable for water-based fluids [3,25].

Polymer-based fluids have several limitations like degradation at a higher temperature, polymer deposition/residue, blockage of pore throat, permeability reduction, fluid leak-off, friction, lower conductivity of the proppant and hydraulic fractures, formation damage and high treatment cost [16,17,25,26,37,41,49,51,60–63]. Additionally, breakers additives are essential for breaking the high viscosity polymer residue and efficiently cleaning up the filter cake after the fracturing process. The limitations of the cross-linked polymer fluids as addressed in several studies showed thermal sensitivity, inefficient transport of the proppant, lower conductivity of fracture and formation damage [16,25,64,65].

The success of viscoelastic surfactant (VES) fracturing fluid is restricted to temperature till 200°F and beyond this range, the viscosity of the VES fluid reduces [3,40,49]. The high surfactant cost, leak-off issues, inability to form a filter cake, ineffective emulsification and wettability alteration throughout the reservoirs and high cleaning time are the other factors that hinder the role of VES fracturing fluid [33,34,37,66,67].

The field application for foam-based fracturing fluids is restricted because of their instability, fracturing interruption at a high shear rate and temperature, friction, blockage of pores by large size droplets formed by foam coalescence [3,57,68].

7.4 Nanotechnology in Unconventional Reservoirs

Nanotechnology plays an important role in terms of hydrocarbon recovery when implemented in unconventional reservoirs. It has the potential to address the limitation and challenges encountered with fracturing fluids. The induction of nanomaterial in the hydraulic fracturing fluid to improve rheology properties for effective fracturing stimulation is one of the most important features. Other advantages include fluid loss control, breaker system, formation fine control, drilling and completion and nano-sensor.

7.4.1 Nanoparticles for Hydraulic Fracturing

The recovery of hydrocarbon from tight reservoirs depends widely on the hydraulic fracturing stimulation, therefore the development of fracturing fluid for reservoirs under different conditions is one of the important criteria for successful fracturing. The applications of nanoparticles towards the development of polymer, surfactant and foam-based fracturing fluid are focussed on and discussed in this section.

7.4.1.1 Nanoparticles in a Polymer-Based Fracturing Fluid

Bio-polymer guar has been extensively used to enhance the viscosity of the fracturing fluid but due to its thermal instability, synthetic polymers such as acrylamide are preferred. These synthetic polymers are loaded with a higher concentration to achieve stable viscosity under a high-temperature scenario. Moreover, higher polymer concentration results in residue, which has the potential to cause severe formation damage. Therefore, nanotechnology can play a major role in such condition to overcome the technical limitation by improving the properties of the fracturing fluid.

Nanoparticles can be used to enhance the viscosity of the polymer-based fracturing fluid. Studies reported that the addition of SiO_2 nanoparticles (20nm diameter) on a guar solution of 33 lb/Mgal can enhance the viscosity of the solution and as the concentration of nanoparticles increases from

0.058 to 0.4 wt%. They reported that the viscosity improvement was due to the formation of supermicellar aggregates by the adsorption of guar on nanoparticles. However, the rheological properties were hammered as the temperature and shear rate of the system increased due to a reduction in the entanglement formed by the stretching of the three-dimensional structure and the Brownian motion of the nanoparticles [69]. In some investigations, SiO_2 nanoparticles were found to even reduce the fluid leak-off issue though the viscosity enhancement may not be large (123–179 cp) [70].

7.4.1.1.1 Nanoparticles in a Cross-Linked Polymer Fluid

A cross-linking agent is used to form a high viscoelastic gel fluid by combining with two or more polymeric chain [16]. This process reduces the polymer loading and is found effective for higher hydrocarbon recovery when surveyed in more than 200 hydro-fractured formation wells [71,72]. The mechanisms as studied for titanium and zirconium complexes clearly show that the complexes initially hydrolyses in the fracturing fluid, which is then condensed into respective nanoparticles (TiO_2 and ZrO_2) and then undergoes cross-linking [58,73]. The parameters that decide the cross-linking activity are the pH and size of the nanoparticles. If optimum pH and size are not maintained, nanoparticles start to agglomerate and become unstable and lower the available surface, which therefore weakens the cross-linking process [16,58].

The sustainability of rheological properties achieved using general cross-linker agents like boron ion, titanium and zirconium complexes with guar gum polymeric-based fluid is challenging as the temperature and pressure of the system enhance [16,74–78]. Therefore to overcome the technical challenges in the conventional cross-linked process, nanoparticles are introduced in the cross-linker (nano-crosslinkers).

7.4.1.1.2 Nanoparticles as Cross-Linkers (Nano-Crosslinkers)

Researchers have developed gels that are stable at high pressure and thus do not give up their rheological properties. The use of boronic acid functionalizes SiO_2 nanoparticles (15 nm diameter) as a nano-crosslinker on guar fluid to form a stable gel. The concentration of the boron used in this process of gel formation is much lesser (7.5 ppm) than that required for a conventional borate cross-linked gel (120 ppm). The reduction in polymer loading along with a lower concentration of borate together in the presence of nanoparticles could initiate the cross-link and improve the rheological properties [79]. The reported data reveal that as the pressure increases to 8000 psi, the viscosity of the gel can be reduced by about 97% without nanoparticles. However, in the presence of nanoparticles, the viscosity can be improved to around six to nine times due to robust crosslink [16]. The rheological properties of the fracturing fluid, which involve nano-crosslinkers, are superior to that of pure nanoparticles [80]. Also, the thermal stability of the gel formed using nano-crosslinkers is superior to that of conventional crosslinkers gel. A study [60] reported that as the temperature of the system increases from 30°C to 75°C, the viscosity retained by the nano-crosslinked gel was beyond 400 mPa s, whereas for the same system, the viscosity reduces to around 50 mPa s in the case of a conventional cross-linked gel.

Additionally, other attractive properties of nano-crosslinkers are negligible polymer cake deposition, better conductivity for hydrocarbon flow, minimum formation damage, effective cleanup and low operation cost [80,81]. Apart from the available literature, some gaps need to be addressed to understand the mechanisms of nano-crosslinker gel formation. The scope could be the studies on Brownian forces, the interaction between the particles, fluid–particle interaction and the viscous forces [3,21].

7.4.1.2 Nanoparticles in a Surfactant-Based Fracturing Fluid

Surfactant at the critical micelle concentration tends to aggregate to form micelles of different shapes like spherical, rod-like or worm-like structures (Figure 7.2) and is held responsible for the viscoelastic properties [49]. The formation damage caused by a VES fluid is less than that of a polymer-based fracturing fluid. The challenges with the surfactant (VES) based fluid includes high leak-off, instability at higher temperature and pressure (shear rate) and screening out of proppant under such conditions [19,62,82–85]. These challenges limit the application of VES-based fluids for reservoirs in which the permeability is <100 to 200 mD approximately [86,87]. Additionally, the operating cost is higher for high permeability reservoirs due to high leak-off. Therefore, despite several advantages (i.e., low molecular weight, minimum formation damage, no filter cake formation, enhanced viscosity, effective proppant transportation, improved clean-up and independent of cross-linking), the nanomaterial is preferred to deploy in a VES fracturing fluid considering their technical limitations [17,48,66,88].

The early articles on a nanoparticles-VES based fluid were successful to maintain a viscosity of around 200 cp for more than 90 minutes at a temperature of 250°F and a shear rate of 100 per seconds, whereas in the absence of nanoparticles (VES solution only), the viscosity of the fluid dropped to around 40 cp. They also observed that the rate of fluid leak-off was slower with nanoparticles-induced VES fluid as it forms pseudo-filter cake, which restricts the fluid leakage in the formation [62]. The nanoparticles in the VES fluid tend to self-aggregate inside the surfactant micelle (lamella or rod/worm-like structure) enhancing the 3D interlink network. The pseudo-crosslink of the nanoparticles with the surfactant structures depends on the chemisorption and electrostatic/surface charge interaction between the nanoparticles and the surfactant molecules [89]. This enhanced viscosity due to pseudo-crosslink of nanoparticles and VES fluid improves the transporting capacity of the proppant even at 110°C [89,90]. However, if nanoparticles exceed their optimum concentration, aggregation occurs which can severely affect the viscosity of the nanoparticles-VES based fracturing fluid [19]. Other factors on which the efficiency of nanoparticles-VES based fluid depends are the concentration of VES solution, the shape of nanoparticles, VES adsorption on nanoparticles, crosslink-like junction formed in between nanoparticles and micelle, salt concentration, and working temperature [17,67].

The success of this type of system depends on the choice of surfactant. Therefore, in order to predict the success of this system, more fundamental or experimental data are required to predict the consistency of the process. The role of different parameters such as nanoparticles size and loading, surfactant concentration, the surface charge of surfactant and nanoparticles, rheological properties to maintain effective proppant transport and temperature/shear resistance dependency needs to be scrutinized further in detail [3,17].

7.4.1.3 Nanoparticles in a Foam-Based Fracturing Fluid

Foam-based fracturing fluids used in the field applications face certain issues like adsorption of surfactant molecules, degradation of surfactants/polymers under extreme reservoir conditions and high investment for chemicals [55]. In addition, the stability of the foam is challenging, the decay rate depends on bubble coalescence, Ostwald ripening and liquid drainage as encountered due to pressure differences [3]. Therefore, in order to enhance the stability as well as rheological properties of the foam-based fluid, nanoparticles are blended. The addition of nanoparticles in the foam-based solution results in the adsorption of nanoparticle at the gas–liquid interface (foam produced), which aggregates to form a network [91–95]. The formation of these particle aggregation networks acts as a barrier towards film thinning, bubble coalescence and liquid drainage thus improving the stability of the rheological properties of a foam-based fluid [96,97].

The investigations performed on nanoparticles–polymer foam-based fluid stated that the addition of nanoparticles can reduce the mobility of injected gas and improve the foam stability from moderate temperature (122°F) to high temperature (250°F) [55,98]. A system that used SiO_2 as nanoparticles, CO_2 as injected gas, guar-gum as polymer and AOS as a foaming agent was successful in enhancing the mechanisms responsible for foam stability and improved viscosity at a higher temperature [55].

Nanoparticles in surfactant foaming dispersion can improve the rheology of the foam, lower fluid loss rate and enhance the proppant transport capacity of the foam [99]. The stability and rheological behaviour of VES and worm-like surfactant structure (micelles) foam obtained by surfactant mixtures in the existence of nanoparticles have been examined [100] and the mechanisms responsible for such a system is the enhancement in the viscosity of the foam fluid achieved by the transformation of the spherical micelles into worm-like micelles [101]. In work-like micelles foam, the nanoparticles actively participate in the networking and act as a pseudo-cross-link agent [101–104]. The apparent viscosity is improved by the formation of a micelle–nanoparticles junction which enhances the cross-link aggregation and actively interlinks the micelle chain resulting in additional viscoelastic properties. The effect of cross-link is time-dependent and after a certain duration (30 minutes as per data reported), the particles tend to segregate from the micelle–nanoparticles junction thus reverting back to the initial fluid viscosity [101,105]. Therefore because of this property, the work-like micelle–nanoparticles foam breaks down, which assists in transportation, improved conductivity with minimum formation damage. However, more experimental findings are required to understand the breaking mechanisms of the micelle–nanoparticles junction and its relation towards effective hydraulic fracturing for field applications under extreme reservoir conditions [3]. Studies on the rheological properties of polymer–VES–nanoparticles-based fracturing fluid as a function of concentration and temperature are also reported [17,69].

7.4.2 Nanoparticles Impact on a Proppant

The condition for effective hydrocarbon flow during hydraulic fracturing stimulation in unconventional reservoirs depends on the opening of the induced micro and nano-fractures after the fluid pressure is released. The proppant is responsible for holding the openings of the bigger or natural fractures after the completion of the fracturing task, thus improving the permeability and porous flow of the hydrocarbon [106]. However, the proppant due to their large size cannot restrict the closure of the tiny micro and nano-sized fractures openings.

An ideal proppant must possess high resistance towards closure stress, high surface area, high sphericity and high elasticity and hardness [106]. These optimum parameters can be induced in conventional proppants by introducing nanoparticles and thus the concept of nano-proppant was invented. Studies reported that nanoparticles and proppant synergy (nano-proppant) can effectively enhance the conductivity of micro and nano-size complex fracture network [107]. The introduction of conventional proppant after nano-proppant can prevent the closure of small/micro/nano fractures opening during and after completion of the fracturing task, which therefore enhances the conductivity. The sphericity of the nanoparticles is responsible for dealing with the high closure stress as compared to the no-spherical proppant [3,107]. The development of a nano-proppant by the application of carbon nanotube [108,109] and nanoparticles stabilized emulsion have also been reported [6]. Thus, the effectiveness of a nano-proppant to transport in the deeper and small-size complex fractures, which are not assessable by conventional proppant, clearly indicates the dominant behaviour of a nano-proppant [6].

The current state-of-the-art does not provide inside of nano-proppant averting the closure of small micro and nano-size fracture openings due to limited investigations. The cheap nano-proppant retaining all the desired properties needs to be developed using different nanoparticles and their properties. Moreover, the development of a nano-proppant should consider different characteristics like thermal stability, chemical inertness, eco-friendly nature, availability, and effective clean-up. Furthermore, more experimental findings are required under reservoir conditions followed by simulation studies incorporating pressure loss at the fractures in order to understand the complexities of the system [3].

7.4.3 Nanoparticles as a Fluid Loss Control Agent

The loss of fluid into the rock matrix during fracturing is one of the severe issues that can damage the formation, impart the conductivity for hydrocarbon flow, shift in the capillary pressure and water blockage [110]. The rate of fluid loss is being controlled by the use of additives, which seals the formation of wall openings during the progress of fracturing [15]. Moreover, the additives are not effective to control the fluid loss in nano-size and micro size fissures because of their larger size with respect to the pore throats. Therefore to overcome this undesirable behaviour, different types of nanoparticles are employed such that the fluid loss from micro and nano-size fractures can be controlled [111,112]. Nanoparticles due to their size can block the small (micro or nano) fractures opening thereby controlling the fluid loss. The effective diameter of the particles deployed as additives for fluid control can be up to one-third of the formation pore throat [15]. The particle sizes from 3 to 300 nm are usually considered ideal for shale formations of pore throat size from 10 nm to 1 mm [15,113–115]. The control on fluid loss depends on the concentration of nanoparticles. The fluid loss decreases with an increase in nanoparticles concentration [111]. Furthermore, nanomaterials can be also used to prepare smart fluids which can establish mud cake (thin and tight) and improve the filtration and rheological properties of conventional hydraulic fracturing fluid [116,117].

7.4.4 Nanoparticles in Sensors

In the petroleum industry, the information obtained by NMR and seismic logging is not suitable for unconventional reservoirs [118]. The information about proppant activity or transportation and hydraulic fracturing is crucial to maximize hydrocarbon flow and its recovery from unconventional reservoirs. Therefore to acquire overall information from such low porous and permeable formation, nanosensors are preferred due to their small sizes which can transport throughout the reservoir fractures [119].

The paramagnetic properties of the nanoparticle can be exploited to gather information about the in-situ characteristics of fluid and rock in addition to the dynamic flow properties on injected fluid during hydraulic fracturing [120]. The adjustment in the magnetic field of nanoparticles can be deployed to improve the magnetic susceptibility [121]. Therefore, some attempts are carried out to use nanoparticles with higher magnetic susceptibility considering the reservoir environment [122]. When a mixture of super paramagnetic nanoparticles and proppant is injected into the well during fracturing, the signal from the downhole/borehole (information required for characterization) enhances and the presence of proppant in the fractures can be identified [119,122,123]. The formulation of optimum conditions for the nano-sensors to travel through the pores/fractures depends on several parameters like nanoparticles shape, particle size, particle distribution and rock material surface charge [124]. Several mathematical models have been developed to explain the dynamic transport of nano-sensors in unconventional reservoirs [121]. However, more accurate numerical models are required, which can be developed by considering the flow dynamic behaviour and different reservoir conditions.

7.4.5 Nanoparticles in Unconventional Gas Reservoirs

The recovery of gas from unconventional reservoirs can be enhanced by hydraulic fracturing stimulation. The capillary effect in such low permeability formation is dominant with respect to the viscous force. The fluid leak-off that occurred during the fracturing process together with the capillary end effect can enhance the water saturation in the vicinity of the fracture face or wellbore resulting in water blockage and thus limits the production of gas [125–128]. The saturation of water near the wellbore is dependent on the wettability of the rock surface [126,129,130]. Therefore, wettability alteration is an effective technique that can reduce water saturation [131,132]. Numerous studies conveyed that wettability alteration can be obtained by transforming the formation rock surface towards a more hydrophobic character by the application of surfactants [126,133,134] and brines (salinity of different compositions) [135,136], which weaken the capillary pressure and declines water blockage. The surfactants may not be suitable at a higher temperature and due to thermal instability, the performance in terms of wettability alteration can be seriously affected. Hence to overcome this limitation, the applications of nanoparticles were introduced.

The investigation on wettability alteration of the grain surface has been studied by several researchers [137–139]. Hydrophobic nanoparticle solution, when interacts with a solid porous surface (adsorbate), undergoes adsorption forming a nano-film on the surface which changes the wettability of the system from hydrophilic to hydrophobic [61,140]. A recent study showed how the change in wettability of tight sand cores was altered from strong liquid wet to a gas wet system using nanoparticles [141].

The alteration in wettability of a system towards neutral or gas wet for improved gas recovery in unconventional reservoirs depends on the capillary–viscous ratio which can be confirmed by further studies. The mechanism of interfacial tension reduction and wettability alteration of the surface are responsible for higher hydrocarbon recovery in conventional reservoirs. Similar behaviour in unconventional reservoirs is yet to develop, which requires further experimental investigation [3]. Nanotechnology applications in the oil and gas industry can further be extended in the area of drilling and completion, tackle the harsh down-hole condition and flow assurance [142,143].

7.5 Field Applications and Challenges in Unconventional Reservoirs

The promoted hydrocarbon recovery by the application of nanotechnology in unconventional reservoirs for field application is not yet reported as per the state-of-art literature. However, few developed commercial studies have been reported [3,22]. The application of nanotechnology as fluid mobility modifiers in the field of Woodford shale play in Oklahoma by Halliburton as commercialization partners had reported an increase in oil recovery (309%–559%) after 30 days of the treatment. Nano-proppants deployed in the fields of South America (gravel pack material), China (coal bed methane) and Europe (Shale and coal bed methane) with Sun drilling technologies as the commercialization partner. Formation fine control additives and clay stabilizers developed from nanotechnology for sand control and fine migration in the field of deep water Gulf of Mexico with Baker Hughes as commercialization partner.

Numerous limitations are encountered which have to be overcome for commercializing the application of nanomaterials in a real field of unconventional reservoirs. One of the major issues with nanoparticles is that it undergoes agglomeration when suspended in fluid. This agglomeration retards the stability of the particles in solutions and drastically reduces the efficiency of the processes. The overall size of the particles enhances due to agglomeration, which blocks the reservoir pores and thus demands higher injection pressure [61]. The economy is the most important factor that decides the applicability of nanoparticles in unconventional reservoirs. Therefore, to overcome the cost impact, nanoparticles can be prepared from cheap sources like fly ash and waste effluent. Moreover, the particles should be eco-friendly and should not result in hazards or health risks. Till date, there are limited experimental and simulation studies that cannot decide the applicability of nanoparticles as fluid loss control additives, nano-proppant and nanosensors in the real oil field. Hence, more rigorous investigations are required to fill the lacuna for the application of nanotechnology in unconventional reservoirs at a large scale.

References

1. G.E. King, and D. Durham, Chater 1 - Environmental Aspects of Hydraulic Fracturing: What Are the Facts? *Hydraulic Fracturing: Environmental Issues, ACS Symposium Series*, vol. 1216, pp. 1-44, 2015, doi:10.1021/bk-2015-1216.ch001.
2. G. M. Hamada, Comprehensive evaluation and development of unconventional hydrocarbon reserves as energy resource, *Archives of Petroleum & Environmental Biotechnology* vol. APEB-102, pp. 1–11, 2016.

3. N. Yekeen, E. Padmanabhan, A. K. Idris, and P. S. Chauhan, Nanoparticles applications for hydraulic fracturing of unconventional reservoirs: A comprehensive review of recent advances and prospects, *Journal of Petroleum Science & Engineering,* vol. 178, pp. 41–73, 2019.
4. G. Kubala, Technology focus: Tight reservoirs *Journal of Petroleum Technology,* vol. 65, p. 124, 2013, 124. doi: 10.2118/1013-0124-JPT.
5. S. Steiner, A. A. Ahsan, A. Noufal, B. F. Franco, and T. Koksalan, Integrated approach to evaluate unconventional and tight reservoirs in Abu Dhabi, SPE-177610, in *Abu Dhabi International Petroleum Exhibition and Conference,* 9–12 November, Abu Dhabi, UAE, 2015.
6. D. Liu, Y. Yan, G. Bai, Y. Yuan, T. Zhu, F. Zhang, et al., Mechanisms for stabilizing and supporting shale fractures with nanoparticles in Pickering emulsion, *Journal of Petroleum Science & Engineering,* vol. 164, pp. 103–109, 2018.
7. L. Britt, C. Hager, and J. Thompson, Hydraulic fracturing in a naturally fractured reservoir, SPE-28717-MS, in *InternationalPetroleumConferenceandExhibitionof Mexico,* Veracruz, Mexico, 10–13 October, 1994. doi: 10.2118/ 28717-MS.
8. A. M. Gomaa, Q. Qu, R. Maharidge, S. Nelson, and T. Reed, New insights into hydraulic fracturing of shale formations, in *IPTC-17594-MS, International Petroleum Technology Conference,* Doha, Qatar, 2014.
9. T. Tella, Estimating reserves forunconventional shale resource plays. *SPEE Presentation,* Tulsa, August 11, 2011.
10. L. Wang, Y. Tian, X. Yu, C. Wang, B. Yao, S. Wang, et al., Advances in improved/ enhanced oil recovery technologies for tight and shale reservoirs *Fuel,* vol. 210, pp. 425–445, 2017.
11. EIA, Annual energy outlook, 2017.
12. G. E. King, Hydraulic fracturing 101: What every representative, environmentalist, regulator, reporter, investor, university researcher, neighbor, and engineer should know about hydraulic fracturing risk, *Journal of Petroleum Technology,* vol. 64, pp. 34–42, 2012.
13. F. Nath and C. Xiao, Characterizing foam: Based frac fluid using carreau rheological model to investigate the fracture propagation and proppant transport in eagleford shale formation, SPE-187527-MS, in *SPE Eastern Regional Meeting,* Lexington, Kentucky, 4–6 October, 2017. doi: 10.2118/187527-MS.
14. T. Martin and M. Economides, *Modern Fracturing Enhancing Natural Gas Production.* ET Publishing, Houston, TX, 2007.
15. R. Barati, Application of nanoparticles as fluid loss control additives for hydraulic fracturing of tight and ultra-tight hydrocarbon-bearing formations, *Journal of Natural Gas Science and Engineering,* vol. 27, pp. 1321–1327, 2015.
16. A. Alharbi, F. Liang, G. A. Al-Muntasheri, and L. Li, Nanomaterials-enhanced high pressure tolerance of borate-crosslinked guar gels, SPE-188817-MS, in *International Petroleum Exhibition & Conference,* 13–16 November, Abu Dhabi, UAE, 2017.
17. G. A. Al-Muntasheri, F. Liang, and K. L. Hull, Nanoparticle-enhanced hydraulic-fracturing fluids: A review, SPE-185161, *SPE Production & Operations,* vol. 32, 186–195, 2016.
18. A. Guerfi, S. Sevigny, M. Lagace, P. Hovington, K. Kinoshita, and K. Zaghib, Nanoparticle $Li_4Ti_5O_{12}$ spinel as electrode for electrochemical generators, *Journal of Power Sources,* vol. 119, pp. 88–94, 2003.

19. M. Zhao, Y. Zhang, C. Zou, C. Dai, M. Gao, Y. Li, et al., Can more nanoparticles induce larger viscosities of nanoparticle: Enhanced wormlike micellar system (NEWMS)? *Materials,* vol. 10, p. 1096, 2017.
20. P. Liu, S. Guo, M. Lian, X. Li, and Z. Zhang, Improving water-injection performance of quartz sand proppant by surface modification with surface-modified nanosilica, *Colloids Surfaces A Physicochemical and Engineering Aspects,* vol. 470, pp. 114–119, 2015.
21. G. Chauhan, A. Verma, A. Hazarika, and K. Ojha, Rheological, structural and morphological studies of Gum Tragacanth and its inorganic SiO_2 nanocomposite for fracturing fluid application., *Journal of the Taiwan Institute of Chemical Engineers,* vol. 80, pp. 978–988, 2017.
22. S. Gottardo, A. Mech, M. Gavriel, C. Gaillard, and B. Sokull-Klüttgen, Use of nanomaterials in fluids, proppants, and downhole tools for hydraulic fracturing of unconventional hydrocarbon reservoirs, JRC Technical Reports, Publications Office of the European Union, Luxembourg, pp. 1–81, 2016.
23. S. Gottardo, V. Amenta, A. Mech, and B. Sokull-Klüttgen, Assessment of the use of substances in hydraulic fracturing of shale gas reservoirs under REACH, September 2013, JRC Scientific and Policy Report, EUR 26069 EN, 2013.
24. A. A. Ghaithan, A critical review of hydraulic fracturing fluid over the last decade, SPE 169552, in *SPE Western North America and Rocky Mountain Joint Regional Meeting,* Denver, Colorado, 16–18 April, 2014.
25. R. Barati and J. T. Liang, A review of fracturing fluid systems used for hydraulic fracturing of oil and gas wells, *Journal of Applied Polymer Science,* vol. 131, pp. 1–11, 2014.
26. L. Li, S. Ozden, G. A. Al-Muntasheri, and F. Liang, Nanomaterials-enhancedhydrocarbon-basedwelltreatmentfluids, SPE-189960-MS, in *SPE International Conference and Exhibitionon Formation Damage Control,* 7–9 February, Lafayette, Louisiana, 2018. doi: 10.2118/189960-MS.
27. L. Gandossi and E. U. Von, An overview of hydraulic fracturing and other formation stimulation technologies for shale gas production, Update 2015, JRC Science for Policy Report, EUR 26347 EN, 2015.
28. J. W. Ely, *Fracturing Fluid and Additives,* SPE Henry L Doherty Monograph Series, SPE, Richardso, TX, 1989.
29. J. Holtsclaw and G. P. Funkhouser, A crosslinkable synthetic: Polymer system for high temperature hydraulic fracturing applications, *SPE Drilling and Completion,* vol. 25, pp. 555–563, 2010.
30. D. V. S. Gupta and P. Carman, Fracturing fluid for extreme temperature conditions is just as easy as the rest, SPE 140176, in *SPR Hydraulic Fracturing Technology Conference and Exhibition,* The Woodlands, TX, 24–26 January.
31. N. Gaillard, A. Thomas, and C. Favero, Novel associative acrylamide-based polymer for proppant transport in hydraulic fracturing fluids, SPE 164072, in *SPE International Symposium on Oilfield Chemistry,* The Woodlands, Texas, 08–10 April, 2013.
32. W. Hurst, Establishment of the skin effect and its impediment to fluid flow in to a wellbore, *Petroleum Engineering* vol. 25, pp. 36–38, B6 through B16, 1953.
33. H. Wu, Q. Zhou, D. Xu, R. Sun, P. Zhang, B. Bai, et al., SiO_2 nanoparticle assisted low-concentration viscoelastic cationic surfactant fracturing fluid, *Journal of Molecular Liquids,* vol. 266, pp. 864–869, 2018.

34. W. F. Pu, D. J. Du, and R. Liu, Preparation and evaluation of supramolecular fracturing fluid of hydrophobically associative polymer and viscoelastic surfactant, *Journal of Petroleum Science & Engineering,* vol. 167, pp. 568–576, 2018.
35. L. Qiu, Y. Shen, T. Wang, and C. Wang, Rheological and fracturing characteristics of a novel sulfonated hydroxypropyl guar gum, *International Journal of Biological Macromolecules,* vol. 117, pp. 974–982, 2018.
36. S. Longo and V. Di Federico, Unsteady flow of shear-thinning fluids in porous media withpressure-dependent properties, *Transport Porous Media,* vol. 110, pp. 429–447, 2015.
37. A. Das, G. Chauhan, A. Verma, P. Kalita, and K. Ojha, Rheological and breaking studies of a novel single-phase surfactant-polymeric gel system for hydraulic fracturing application, *Journal of Petroleum Science & Engineering,* vol. 167, pp. 559–567, 2018.
38. J. Song, W. Fan, X. Long, L. Zhou, C. Wang, and G. Li, Rheological behaviors of fluorinated hydrophobically associating cationic guar gum fracturing gel, *Journal of Petroleum Science & Engineering,* vol. 146, pp. 999–1005, 2016.
39. H. Sun and Q. Qu, High - efficiency boron crosslinkers for low: Polymer fluids, SPE 140817, in *SPE International Symposium on Oilfield Chemistry,* The Woodlands, Texas, 11–13 April, 2011.
40. D. Gupta, Unconventional fracturing fluids for tight gas reservoirs, SPE 119424, in *SPE Hydraulic Fracturing Technology Conference,* The Woodlands, TX, 19–21 January, 2009. doi: 10.2118/119424-MS.
41. Y. Zhang, C. Dai, Y. Qian, X. Fan, J. Jiang, Y. Wu, et al., Rheological properties and formation dynamic filtration damage evaluation of a novel nanoparticle-enhanced VES fracturing system constructed with wormlike micelles, *Colloids Surfaces A: Physicochemical and Engineering Aspects,* vol. 553, pp. 244–252, 2018.
42. A. S. Teot, M. Ramaiah, and M. D. Coffey, Aqueous wellbore service fluids, U. S. Patent No. 4,725,372, 1988.
43. K. Zhang, Fluids for fracturing subterranean formations, U. S. Patent No. 6,468,945, 2002.
44. G. F. di Lullo Arias, P. Rae, and A. J. K. Ahmad, Viscous fluid applicable for treating subterranean formations, U. S. Patent No. 6,491,099, 2002.
45. M. S. Dahanayake, J. Yang, J. H. Y. Niu, P. J. Derian, R. Li, and D. Dino, Viscoelastic surfactant fluids and related methods of use, U. S. Patent No. 6,831,108, 2004.
46. R. D. Koehler, S. R. Raghavan, and E. W. Kaler, Micro structure and dynamics of wormlike micellar solutions formed by mixing cationic and anionic surfactants, *The Journal of Physical Chemistry B,* vol. 104, pp. 11035–11044, 2000.
47. D. Angelescu, A. Khan, and H. Caldararu, Viscoelastic properties of sodium-dodecyl sulfate with aluminum salt in aqueous solution, *Langmuir,* vol. 19, pp. 9155–9161, 2003.
48. G. Chauhan, K. Ojha, and A. Baruah, Effects of nanoparticles and surfactant charge groups on the properties of VES gel, *Brazilian Journal of Chemical Engineering,* vol. 34, pp. 241–251, 2017.
49. M. R. Gurluk, H. A. Nasr-El-Din, and J. B. Crews, Enhancing the performance of viscoelastic surfactant fluids using nanoparticles, SPE-164900, in *EAGE Annual Conference & Exhibition Incorporating,* 10–13 June, SPE Europec, London, UK, 2013. doi: 10.2118/164900-MS.

50. A. Baruah, D. S. Shekhawat, A. K. Pathak, and K. Ojha, Experimental investigation of rheological properties in zwitterionic-anionic mixed-surfactant based fracturing fluids, *Journal of Petroleum Science & Engineering,* vol. 146, pp. 340–349, 2016.
51. Z. Yan, C. Dai, M. Zhao, Y. Sun, and G. Zhao, Development, formation mechanism and performance evaluation of a reusable viscoelastic surfactant fracturing fluid, *Journal of Industrial and Engineering Chemistry,* vol. 37, pp. 115–122, 2016.
52. P. M. Mcelfresh and C. F. Williams, Hydraulic fracturing using non-ionic surfactant gelling agent, in GooglePatents, United Statesed: Google Patents US7,216,709 May 15, Washington, DC, U. S. Patent and Trade mark Office, 2007.
53. R. Anandan, S. Johnson, and R. Barati, Polyelectrolyte complex stabilized CO_2 foam systems for hydraulic fracturing application, SPE-187489-MS, in *SPE Liquids-Rich Basins Conference - North America,* 13–14 September, Midland, Texas, 1–19, 2017. doi: 10.2118/187489-MS.
54. S. Tong, R. Singh, and K. K. Mohanty, Proppant transport in fractures with foam-based fracturing fluids, SPE-187376-MS, in *SPE Annual Technical Conference and Exhibition,* 9–11 October, San Antonio, Texas, 2017. doi: 10.2118/187376-MSUS9845427.
55. A. S. Emrani, A. F. Ibrahim, and H. A. Nasr-El-Din, Mobility control using nanoparticle stabilized CO_2 foam as a hydraulic fracturing fluid, SPE-185863-MS, in *SPE Europec Featured at 79th EAGE Conference and Exhibition,* 12–15 June, Paris, France, 2017. doi: 10.2118/185863-MS.
56. C. Xiao, S. N. Balasubramanian, and L. W. Clapp, Rheology of supercritical CO_2 foam stabilized by nanoparticles, SPE-179621-MS, in *SPE Improved Oil Recovery Conference,* 11–13 April, Tulsa, Oklahoma, 2016. doi: 10.1021/acs.iecr.7b01404.
57. N. Yekeen, E. Padmanabhan, and A. K. Idris, A review of recent advances in foam based fracturing fluid application in unconventional reservoirs, *Journal of Industrial and Engineering Chemistry,* vol. 66, pp. 45–71, 2018.
58. T. Hurnaus and J. Plank, Behavior of titania nanoparticles in cross-linking hydroxy propyl guar used in hydraulic fracturing fluids for oil recovery, *Energy and Fuels,* vol. 29, pp. 3601–3608, 2015.
59. C. H. Bivins, C. Boney, C. Fredd, J. Lassek, P. Sullivan, J. Engels, et al., New fibers for hydraulic fracturing, *Science,* vol. 83, pp. 660–686, 2002.
60. Z. Zhang, H. Pan, P. Liu, M. Zhao, X. Li, and Z. Zhang, Boric acid in corporated on the surface of reactive nanosilica providing a nano-crosslinker with potential in guar gum fracturing fluid, *Journal of Applied Polymer Science,* vol. 134, p.45037, 2017.
61. T. Wang, Y. Zhang, L. Li, Z. Yang, Y. Liu, J. Fang, et al., Experimental study on pressure-decreasing performance and mechanism of nanoparticles in low permeability reservoir, *Journal of Petroleum Science & Engineering,* vol. 166, pp. 693–703, 2018.
62. J. B. Crews and T. Huang, Nanotechnology applications in viscoelastic surfactant stimulation fluids, *SPE Production & Operations,* vol. 23, pp. 512–517, 2008. doi: 10.2118/107728-PA.
63. J. T. Srivatsa and M. B. Ziaja, An experimental investigation on use of nanoparticles as fluid loss additives in a surfactant-polymer based drilling fluids, IPTC14952, in *International Petroleum Technology Conference,* 15–17, November, Bangkok, Thailand, 2011. doi: 10.2523/IPTC-14952-MS.

64. R. Phil and L. Glnodi, *Fracturing Fluids and Breaker Systems: A Review of the State of the Art*, SPE, Houston, TX, p. 37359, 1996.
65. K. Wang, Y. Wang, J. Ren, and C. Dai, Highly efficient nanoboron crosslinker for low-polymer loading fracturing fluid system in *SPE/IATMI Asia Pacific Oil & Gas Conference and Exhibition*, 17–19 October, Jakarta, Indonesia, 2017, pp. 1–11.
66. T. Huang and J. B. Crews, Nanotechnology applications in viscoelastic surfactant stimulation fluids, SPE-107728-PA, *SPE Production & Operations*, vol. 23, pp. 512–517, 2008. doi: 10. 2118/107728-PA.
67. A. Hanafy, F. Najem, and H. A. Nasr-El-Din, Impact of nanoparticles shape on the VES performance for high temperature applications, SPE-190099-MS, in *SPE Western Regional Meeting*, Garden Grove, 22–26 April, California, USA, 2018. doi: 10.2118/190099-MS.
68. P. Kundu, V. Kumar, and I. M. Mishra, Experimental study on flow and rheological behavior of oil-in-water emulsions in unconsolidated porous media: Effect of particle size and phase volume fractions, *Powder Technology*, vol. 348, pp. 821–833, 2019.
69. M. F. Fakoya and S. N. Shah, Rheological properties of surfactant-based and polymeric nano-fluid, SPE -163921, in *SPE/ICoTA Coiled Tubing and Well Intervention Conference and Exhibition*, 26–27 March, The Woodlands, Texas, 2013. doi: 10.2118/163921-MS.
70. C. Vipulanandan, A. Mohammed, and Q. Qu, Characterizing the hydraulic fracturing fluid modified with nano silica proppant, in *AADE Fluids Technical Conference and Exhibition*, 15–16 April, Houstan, AADE-4-FTCE-39, 2014.
71. N. H. Kostenuk and P. J. Gagnon, Polymer reduction leads to increased success: a comparativestudy, SPE-100467-PA, in *SPE GasTechnology Symposium*, 15–17 May, 2006, Calgary, Alberta, Canada, doi: 10.2118/100467-PA, 2006.
72. T. Huang, J. B. Crews, and G. Agrawal, Nanoparticle pseudo crosslinked micellar fluids: optimal solution for fluid-loss control with internal breaking, SPE-128067, in *SPE International Symposium and Exhibitionon Formation Damage Control*, 10–12 February, Lafayette, Louisiana, 2010. doi: 10.2118/128067-MS.
73. T. Hurnaus and J. Plank, Crosslinking of guar and HPG based fracturing fluids using ZrO_2 nanoparticles, SPE-173778-MS, in *SPE International Symposiumon Oil field Chemistry*, 13–15 April, The Woodlands, Texas, 2015. doi: 10.2118/ 173778-MS.
74. K. W. England and M. D. Parris, Viscosity influences of high pressure on borate crosslinked gels, SPE-136187, in *SPE Deepwater Drilling and Completions Conference*, 5–6 October, Galveston, Texas, 2010. doi: 10.2118/136187-MS.
75. M. J. Fuller and K. J. Blake, Implications of pressure-induced thinning in crosslinked fluids for fracturing and frac pack operations, SPE-179147-MS, in *SPE Hydraulic Fracturing Technology Conference*, 9–11 February, TheWoodlands, Texas, 2016. doi: 10.2118/179147-MS201.
76. M. Marquez, N. Tonmukayakul, L. A. Schafer, M. B. Zielinski, P. Lord, and T. L. Goosen, High pressure testing of borate crosslinked fracturing fluids, SPE-152593MS, in *SPE Hydraulic Fracturing Technology Conference*, 6–8 February, The Woodlands, Texas, 2012. doi: 10.2118/152593-MS.
77. M. D. Parris, B. A. MacKay, J. W. Rathke, R. J. Klingler, and R. E. Gerald, Influence of pressure on boron cross-linked polymer gels, *Macromolecules*, vol. 41, pp. 8181–8186, 2008.
78. G. A. Al-Muntasheri, A critical review of hydraulic fracturing fluids for moderate to ultralow-permeability formations over the last decade, SPE-169552-PA, *SPE Production & Operations*, vol. 29, pp. 243–260, 2014.

79. V. Lafitte, G. J. Tustin, B. Drochon, and M. D. Parris, Nanomaterials in fracturing applications, SPE-155533, in *SPE International Oil field Nanotechnology Conference and Exhibition*, 12–14 June, Noordwijk, The Netherlands, 2012. doi: 10.2118/155533-MS.
80. F. Liang, G. Al-Muntasheri, H. Ow, and J. Cox, Reduced-polymer-loading, high temperature fracturing fluids by use of nanocrosslinkers, *Society of Petroleum Engineers Journal*, vol. 22, pp. 622–631, 2017.
81. F. Chen, Y. Yang, J. He, T. Bu, X. He, K. He, et al., The gelation of hydroxypropyl guar gum by nano-ZrO_2, *Polymers for Advanced Technologies*, vol. 29, pp. 587–593, 2018.
82. S. S. Sangaru, P. Yadav, T. Huang, G. Agrawal, and F. F. Chang, Temperature dependent influence of nanoparticles on rheological properties of VES fracturing fluid, SPE186943-MS, in *SPE/IATMI Asia Pacific Oil & Gas Conference and Exhibition*, 17–19 October, Jakarta, Indonesia, 2017. doi: 10.2118/186943-MS.
83. R. G. Shrestha, L. K. Shrestha, and K. Aramaki, Wormlike micelles in mixed amino acid-based anionic/nonionic surfactant systems, *Journal of Colloid and Interface Science*, vol. 322, pp. 596–604, 2008.
84. A. Ponton and D. Quemada, Non linear rheology of wormlike micelles, in Emri, I. (Ed.), *Progress and Trends in Rheology V. Steinkopff*, Springer, Heidelberg, pp. 535–536, 1998. doi: 10.1007/978-3-642-51062-5_260.
85. N. Spenley, M. Cates, and T. Mcleish, Non linear rheology of wormlike micelles, *Physical Review Letters*, vol. 71, p. 939, 1993.
86. P. F. Sullivan, B. R. Gadiyar, R. H. Morales, R. A. Holicek, D. C. L. Sorrells, J, and D. D. Fischer, Optimization of a Visco-Elastic Surfactant (VES) fracturing fluid for application in high-permeability formations, SPE-98338-MS, in *SPE International Symposium and Exhibition on Formation Damage Control*, 15–17 February, Lafayette, Louisiana, 2006. doi: 10.2118/98338-MS.
87. J. B. Crews, T. Huang, and W. R. Wood, New technology improves performance of viscoelastic surfactant fluids, SPE-103118-PA, *SPE Drilling & Completion*, vol. 23, pp. 41–47, 2008. doi: 10.2118/103118-PA.
88. H. Tamayo, K. Lee, and R. Taylor, Enhanced aqueous fracturing fluid recovery from tight gas formations: Foamed CO_2 pre-pad fracturing fluid and more effective surfactant systems, in *PETSOC-2007–112: Canadian International Petroleum Conference*, 12–14 June, Calgary, Alberta, 2007. doi: 10.2118/2007112.
89. J. B. Crews and T. Huang, Internal breakers for viscoelastic surfactant fracturing fluids, SPE-106216-MS, in *International Symposium on Oil field Chemistry*, 28 February to 2 March, Houston, Texas, 2007. doi: 10.2118/106216-MS.
90. A. Baruah, A. K. Pathak, and K. Ojha, Study on rheology and thermal stability of mixed (nonionic–anionic) surfactant based fracturing fluids, *AIChE Journal*, vol. 62, pp. 2177–2187, 2016.
91. T. S. Horozov, Foams and foam films stabilised by solid particles, *Current Opinion in Colloid & Interface Science*, vol. 13, pp. 134–140, 2008.
92. A. L. Fameau and A. Salonen, Effect of particles and aggregated structures on the foam stability and aging, *Comptes Rendus Physique*, vol. 15, pp. 748–760, 2014.
93. S. Kumar and A. Mandal, Investigation on stabilization of CO_2 foam by ionic and nonionic surfactants in presence of different additives for application in enhanced oil recovery, *Applied Surface Science*, vol. 420, pp. 9–20, 2017.

94. N. Yekeen, M. A. Manan, A. K. Idris, E. Padmanabhan, R. Junin, A. M. Samin, et al., A comprehensive review of experimental studies of nanoparticles-stabilized foam for enhanced oil recovery, *Journal of Petroleum Science & Engineering*, vol. 164, pp. 43–74, 2018.
95. A. Alharbi, F. Liang, G. A. Al-Muntasheri, and L. Li, Nanomaterials-enhanced high pressure tolerance of borate-crosslinked guar gels, SPE-188817-MS, in *Abu Dhabi International Petroleum Exhibition & Conference*, 13–16 November, AbuDhabi, UAE, 2017. doi: 10.2118/188817-MS.
96. N. Yekeen, A. K. Idris, M. A. Manan, A. M. Samin, A. R. Risal, and T. X. Kun, Bulk and bubble-scale experimental studies of influence of nanoparticles on foam stability, *Chinese Journal of Chemical Engineering*, vol. 25, pp. 347–357, 2017.
97. N. Yekeen, M. A. Manan, A. K. Idris, A. M. Samin, and A. R. Risal, Experimental investigation of minimizationin surfactant adsorption and improvement in surfactant foam stability in presence of silicon dioxide and aluminum oxide nanoparticles, *Journal of Petroleum Science & Engineering*, vol. 159, pp. 115–134, 2017.
98. M. Prodanovic and K. Johnston, Development of nanoparticle-stabilized foams to improve performance of water-less hydraulic fracturing, in *Mastering the Subsurface through Technology, Innovation, and Collaboration*, US Department of Energy, National Energy Technology Laboratory, Pittsburgh, PA, 2016.
99. Q. Lv, Z. Li, B. Li, S. Li, and Q. Sun, Study of nanoparticle–surfactant-stabilized foam as a fracturing fluid, *Industrial and Engineering Chemistry Research*, vol. 54, pp. 9468–9477, 2015.
100. A. FaridIbrahim and H. Nasr-El-Din, Stability improvement of CO_2 foam for enhanced oil recovery applications using nanoparticles and viscoelastic surfactants, SPE-191251-MS, in *SPE Trinidad and Tobago Section Energy Resources Conference*, 25–26 June, Port of Spain, Trinidad and Tobago, 2018. doi: 10.2118/191251-MS.
101. Y. Fei, J. Zhu, B. Xu, X. Li, M. Gonzalez, and M. Haghighi, Experimental investigation of nanotechnology on worm-like micelles for high-temperature foam stimulation, *Journal of Industrial and Engineering Chemistry*, vol. 50, pp. 190–198, 2017.
102. F. Nettesheim, M. W. Liberatore, T. K. Hodgdon, N. J. Wagner, E. W. Kaler, and M. Vethamuthu, Influence of nanoparticle addition on the properties of wormlike micellar solutions, *Langmuir*, vol. 24, pp. 7718–7726, 2008.
103. R. Bandyopadhyay and A. Sood, Effect of silica colloids on the rheology of viscoelastic gels formed by the surfactant cetyl trimethylammonium tosylate, *Journal of Colloid and Interface Science*, vol. 283, pp. 585–591, 2005.
104. G. A. Gaynanova, A. R. Valiakhmetova, D. A. Kuryashov, N. Y. Bashkirtseva, and L. Y. Zakharova, Mixed systems based on erucyl amidopropyl betaine and nanoparticles: Self-organization and rheology, *Journal of Surfactants and Detergents*, vol. 18, pp. 965–971, 2015.
105. J. Fink, *Hydraulic Fracturing Chemicals and FluidsTechnology*. Gulf Professional Publishing, Waltham, MA, 2013.
106. E. Ottestad, Nano-porous proppants for shale oil and gas production. Master Thesis, NTNU Trondheim, Norwegian University of Science and Technology, 2014. http://www.diva-portal.org/smash/get/diva2:753411/FULLTEXT01.pdf.

107. C. C. Bose, B. Fairchild, T. Jones, A. Gul, and R. B. Ghahfarokhi, Application of nanoproppants for fracture conductivity improvement by reducing fluid loss and packing of micro-fractures, *Journal of Natural Gas Science and Engineering,* vol. 27, pp. 424–431, 2015.
108. A. Hussain, Carbon Nanotubes (CNT) for enhanced oil production from shales, NTNU, Master's Thesis, 2014.
109. W. Qin, L. Yue, G. Liang, G. Jiang, J. Yang, and Y. Liu, Effect of multi-walled carbon nanotubes on linear viscoelastic behavior and microstructure of zwitterionic wormlike micelle at high temperature, *Chemical Engineering Research and Design,* vol. 123, pp. 14–22, 2017.
110. R. Anandan, R. Barati, and S. Johnson, Polyelectrolyte complex stabilized CO_2 foam systems for improved fracture conductivity and reduced fluid loss, SPE-191424, in *18IHFT-MSSPE International Hydraulic Fracturing Technology Conference and Exhibition,* 16–18 October, Muscat, Oman, 2018. doi: 10.2118/19142418IHFT-MS.
111. Q. Lv, Z. Li, B. Li, D. Shi, C. Zhang, and B. Li, Silica nanoparticles as a high performance filtrate reducer for foam fluid in porous media, *Journal of Industrial and Engineering Chemistry,* vol. 45, pp. 171–181, 2017.
112. Q. Lv, Z. Li, B. Li, C. Zhang, D. Shi, C. Zheng, et al., Experimental study on the dynamic filtration control performance of N_2/liquid CO_2 foam in porous media, *Fuel,* vol. 202, pp. 435–445, 2017.
113. G. Li, J. Zhang, and Y. Hou, Nanotechnology to improve sealing ability of drilling fluids for shale with micro-cracks during drilling, SPE-156997, in *SPE International Oil field Nanotechnology Conference and Exhibition,* 12–14 June, Noordwijk, The Netherland, 2012. doi: 10.2118/156997-MS.
114. T. Sensoy, M. E. Chenevert, and M. M. Sharma, Minimizing water invasion in shales using nanoparticles, SPE124429, in *SPE Annual Technical Conference and Exhibition,* 4–7 October, New Orleans, Louisiana, 2009. doi: 10.2118/124429-MS.
115. C. C. Bose, B. Alshatti, L. Swartz, A. Gupta, and R. Barati, Dual application of polyelectrolyte complex nanoparticles as enzyme breaker carriers and fluid loss additives for fracturing fluids, SPE-171571-MS, in *SPE/CSUR Unconventional Resources Conference–Canada,* 30 September–2 October, Calgary, Alberta, Canada, 2014. doi: 10.2118/171571-MS.
116. M. Amanullah, M. K. Alarfaj, and Z. A. Al-Abdullatif, Preliminary test results of nanobased drilling fluids for oil and gas field application, SPE-139534-MS, in *SPE/IADC Drilling Conference and Exhibition,* 1–3 March, Amsterdam, The Netherlands, 2011. doi: 10.2118/139534-MS.
117. M. Fakoya and S. Shah, Enhancement of filtration properties in surfactant-based and polymeric fluids by nanoparticles, SPE-171029-MS, in *SPE Eastern Regional Meeting,* 21–23 October, Charleston, WV, 2014. doi: 10.2118/171029-MS.
118. K. E. Cawiezel and D. Gupta, Successful optimization of viscoelastic foamed fracturing fluids with ultralight weight proppants for ultralow permeability reservoirs, SPE-119626-MS, in *SPE Hydraulic Fracturing Technology Conference,* 19–21 January, The Woodlands, Texas, 2009. doi: 10.2118/119626-MS.
119. L. Morrow, D. K. Potter, and A. R. Barron, Detection of magnetic nanoparticles against proppant and shale reservoir rocks, *Journal of Experimental Nanoscience,* vol. 10, pp. 1028–1041, 2015.

120. H. Chun and C. Heechan, Silica, fly ash and magnetite nanoparticles for improved oil and gas production, *Journal of the Korean Society of Mineral and Energy Resources Engineers,* vol. 55, pp. 272–284, 2018. doi: 10.32390/ksmer.2018.55.4.272.
121. C. An, M. Alfi, B. Yan, and J. E. Killough, A new study of magnetic nanoparticle transport and quantifying magnetization analysis in fractured shale reservoir using numerical modeling, *Journal of Natural Gas Science and Engineering,* vol. 25, pp. 502–521, 2016.
122. L. Morrow, B. Snow, A. Ali, S. J. Maguire-Boyle, Z. Almutairi, D. K. Potter, et al., Temperature dependence on the mass susceptibility and mass magnetization of superparamagnetic Mn–Zn–ferrite nanoparticles as contrast agents for magnetic imaging of oil and gas reservoirs, *Journal of Experimental Nanoscience,* vol. 13, pp. 107–118, 2018.
123. A. A. Aderibigbe, K. Cheng, Z. Heidari, J. E. Killough, T. Fuss, and W. Stephens, Detection of propping agents in fractures using magnetic susceptibility measurements enhanced by magnetic nanoparticles, SPE-170818-MS, in *SPE Annual Technical Conference and Exhibition,* 27–29 October, Amsterdam, The Netherlands, 2014. doi: 10.2118/170818-MS.
124. M. N. Alaskar, M. F. Ames, S. T. Connor, C. Liu, Y. Cui, K. Li, et al., Nanoparticle and microparticle flow in porous and fractured media–an experimental study, SPE-146752-PA, *SPE Journal,* vol. 17, pp. 160–161, 2012.
125. S. M. Odumabo and Z. T. Karpyn, Investigation of gas flow hindrance due to fracturing fluid leak off in low permeability sandstones, *Journal of Natural Gas Science and Engineering,* vol. 17, pp. 1–12, 2014.
126. S. Naik, Z. You, and P. Bedrikovetsky, Effect of wettability alteration on productivity enhancement in unconventional gas reservoirs: application of nanotechnology, SPE177021-MS, in *SPE Asia Pacific Unconventional Resources Conference and Exhibition,* 9–11 November, Brisbane, Australia, 2015. doi: 10.2118/177021-MS.
127. X. Xie, Y. Liu, M. Sharma, and W. W. Weiss, Wettability alteration to increase deliverability of gas production wells, *Journal of Natural Gas Science and Engineering,* vol. 1, pp. 39–45, 2009.
128. H. Bahrami, M. R. Rezaee, D. H. Nazhat, J. Ostojic, M. B. Clennell, and A. Jamili, Effect of water blocking damage on flow efficiency and productivity in tight gas reservoirs, SPE-142283, in *SPE Production and Operations Symposium,* 27–29 March, Oklahoma City, Oklahoma, 2011. doi: 10.2118/142283-MS.
129. G. Li, W. Ren, Y. Meng, C. Wang, and N. Wei, Micro-flow kinetics research on water invasion in tight sandstone reservoirs, *Journal of Natural Gas Science and Engineering,* vol. 20, pp. 184–191, 2014.
130. X. Yang, Y. Meng, X. Shi, and G. Li, Influence of porosity and permeability heterogeneity on liquid invasion in tight gas reservoirs, *Journal of Natural Gas Science and Engineering,* vol. 37, pp. 169–177, 2017.
131. W. G. Anderson, Wettability literature survey part 5: The effects of wettability on relative permeability, *Journal of Petroleum Technology,* vol. 39, pp. 1453–1468, 1987.
132. W. G. F. Ford, G. S. Penny, and J. E. Briscoe, Enhanced water recovery improves stimulation results, *SPE Production Engineering,* vol. 3, pp. 515–521, 1988.
133. K. Kumar, E. K. Dao, and K. K. Mohanty, Atomic force microscopy study of wettability alteration by surfactants, *SPE Journal,* vol. 13, pp. 137–145, 2008.

134. P. Chen and K. K. Mohanty, Wettability alteration in high temperature carbonate reservoirs, SPE-169125, in *SPE Improved Oil Recovery Symposium*, 12–16 April, Tulsa, Oklahoma, 2014.
135. D. Lin, X. Tian, F. Wu, and B. Xing, Fate and transport of engineered nanomaterials in the environmental, *Journal of Environmental Quality*, vol. 39, pp. 1896–1908, 2010.
136. R. A. Nasralla, M. A. Bataweel, and H. A. Nasr-El-Din, Investigation of wettability alteration and oil recovery improvement by low salinity water in sandstone rock, *Journal of Canadian Pelroleum Technology*, vol. 52, pp. 144–154, 2013.
137. A. Maghzi, A. Mohebbi, R. Kharrat, and M. H. Ghazanfari, Pore-scale monitoring of wettability alteration by silica nanoparticles during polymer flooding to heavy oil in a five-spot glass micromodel, *Transport in Porous Media*, vol. 87, pp. 653–664, 2011.
138. J. Giraldo, P. Benjumea, S. Lopera, F. B. Cortes, and M. A. Ruiz, Wettability alteration of sandstone cores by alumina-based nanofluids, *Energy Fuels*, vol. 27, pp. 3659–3665, 2013.
139. A. Karimi, Z. Fakhroueian, A. Bahramian, N. P. Khiabani, J. B. Darabad, R. Azin, et al., Wettability Alteration in carbonates using zirconium oxide nanofluids: EOR implications, *Energy Fuels*, vol. 26, pp. 1028–1036, 2012.
140. J. Li, X. Li, K. Wu, D. Feng, T. Zhang, and Y. Zhang, Thickness and stability of water film confined inside nanoslits and nanocapillaries of shale and clay, *International Journal of Coal Geology*, vol. 179, pp. 253–268, 2017.
141. M. Franco-Aguirre, R. Zabala, S. H. Lopera, C. A. Franco, and F. B. Cortés, Interaction of anionic surfactant-nanoparticles for gas-wettability alteration of sandstone in tight gas-condensate reservoirs, *Journal of Natural Gas Science and Engineering*, vol. 51, pp. 53–64, 2018.
142. J. Yang, S. Ji, R. Li, W. Qin, and Y. Lu, Advances of nanotechnologies in oil and gas industries, *Energy Exploration and Exploitation*, vol. 33, pp. 639–657, 2015.
143. M. T. Alsaba, M. F. Dushaishi, and A. K. Abbas, A comprehensive review of nanoparticles applications in the oil and gas industry, *Journal of Petroleum Exploration and Production Technology*, 2020. doi: 10.1007/s13202-019-00825-z.

Index